THEOPHRASTUS
ON STONES

THEOPHRASTUS ON STONES

INTRODUCTION,
GREEK TEXT, ENGLISH TRANSLATION,
AND COMMENTARY

EARLE R. CALEY
THE OHIO STATE UNIVERSITY

JOHN F. C. RICHARDS
COLUMBIA UNIVERSITY

COLUMBUS, OHIO
THE OHIO STATE UNIVERSITY
1956

GRADUATE SCHOOL MONOGRAPHS
CONTRIBUTIONS IN PHYSICAL SCIENCE, NO. 1

Copyright © 1956
The Ohio State University
Columbus

PREFACE

IN spite of its fundamental importance in the history of mineralogy and chemical technology, no modern annotated translation of the treatise *On Stones* by Theophrastus has been available to students of the history of science. Over two hundred years have elapsed since the appearance of the first English translation by John Hill in 1746, and over one hundred and fifty years since the publication of Hill's second edition in 1774. The first French and German versions, which are largely based on Hill's translation, are not only difficult to obtain now but are also obsolete in many ways, especially in their scientific notes. Within the last fifty years a French translation by F. de Mély and a German translation by K. Mieleitner have been published as parts of other works, but they are not accompanied by either text or commentary. We believe that the growing interest in the history of pure and applied science warrants the publication of a new and annotated English translation of this important Greek work. Such a translation is especially desirable at the present time, since few students of science are now able to acquire a reading knowledge of Greek. Even students of Greek who are unfamiliar with the peculiar style and terminology of Theophrastus may find this translation useful. The text and critical notes should also be of interest to them, and the commentary may be of value, for without some interpretation the numerous technical terms used in the treatise and the rationale of the processes described in it are not easily understood. We hope that the occasional items of miscellaneous information scattered through the work may be of some interest to students in other fields. There is ample evidence that this particular treatise has been neglected by scholars generally, and Hill's quaint prefatory remarks are nearly as applicable today as they were over two centuries ago:

> The many References to Theophrastus, and the Quotations from him, so frequent in the Works of all the later Writers of Fossils, would make one believe, at first sight, that nothing was more universally known, or perfectly understood, than the Treatise before us: But when we come to enquire more strictly into the Truth, and examine with our own Eyes what it really is that he has left us, we shall

PREFACE

find that though no Author is so often quoted, no Author is so little understood, or, indeed, has been so little read; those who are so free with his Name, having given themselves, generally, very little Trouble about his Works, and only taken upon Trust from one another, what we shall in most Cases find, on strict Enquiry, to have been originally quoted from him by Pliny . . . .

Many long delays and interruptions have occurred during the preparation of this book, and four authors have been engaged in its composition. In 1934, Earle R. Caley of the Department of Chemistry at Princeton University and Shirley H. Weber of the Department of Classics began work on an annotated translation of the treatise. But for a long time circumstances prevented them from collaborating effectively. In the period between 1936 and the beginning of World War II they lived both in Greece and in the United States and could seldom meet to discuss the work; and though both were in the United States during most of the war, not much attention could be paid to the preparation of the book because of more pressing interests. As a consequence the manuscript was never brought to a satisfactory state, and when Professor Weber returned to Athens at the close of the war to resume his position as Librarian of the Gennadion, he decided that it would be impossible to continue as a collaborator. In the meantime, Thomas T. Read of the School of Mines at Columbia University and John F. C. Richards of the Department of Greek and Latin began, quite independently, to prepare an annotated translation of the same treatise. Early in 1946 they heard that a similar enterprise had long been in progress at Princeton and arranged to collaborate with Earle R. Caley of the Department of Chemistry. After Professor Read died in 1947, the present authors decided to complete the book. We hereby acknowledge our indebtedness to Professor Weber and to Professor Read. Professor Weber gave much time and thought to problems of translation and interpretation; some of his suggestions are incorporated in the present translation and in certain notes in the Commentary. Professor Read was planning to make his own contribution. It is greatly to be regretted that, owing to his illness, he could not use his extensive knowledge of geology and mineralogy for this purpose.

Since both linguistic and scientific knowledge is required, it

PREFACE

seems unlikely that a book of this kind could properly be produced in this age of specialized scholarship without collaboration. In such a book it is not easy to show the exact contribution of each collaborator, but the division of labor and responsibility was approximately as follows. John F. C. Richards collated the manuscripts and editions and prepared the critical notes to the Greek text. He is also responsible for the translation, though Earle R. Caley made many suggestions, so that to some degree this translation may be considered a joint production. The Introduction and the Commentary were written jointly; John F. C. Richards supplied the notes of linguistic interest, but most of the material in the Commentary was contributed by Earle R. Caley. This includes the identification of precious stones and other mineral substances, the discussion of problems in the field of archaeology, chemistry, or mineralogy, and other matters of scientific or technological interest.

We hereby express our sincere thanks to all those who have helped us in our task. Among those who have been especially helpful with criticisms and suggestions are Professor Gilbert Highet of Columbia University and Professors Kenneth M. Abbott, William R. Jones, Lowell Ragatz, and Everett Walters of the Ohio State University. Finally we express our gratitude to the authorities of the Graduate School of the Ohio State University for their generosity in sponsoring the publication of this book.

E. R. C.
J. F. C. R.

COLUMBUS, OHIO
September, 1956

CONTENTS

Introduction	3
Manuscripts and Editions	11
Abbreviations	15
Text	19
Apparatus Criticus	31
Translation	45
Commentary	63
Works Cited in the Commentary	223
Greek Index	227
Index to Translation and Commentary	231

THEOPHRASTUS
ON STONES

INTRODUCTION

PLINY, in his *Natural History*, mentions about twenty Greek writers as authorities for his chapters on precious stones and other mineral substances, but, of the works of these authors, only the brief, or fragmentary, treatise *On Stones* by Theophrastus has survived to inform us in a direct way of the extent of Greek learning in this field. As the earliest known scientific work dealing expressly with minerals and artificial products derived from them, it is of unique importance in the history of mineralogy and of chemical technology.

Theophrastus, the famous pupil of Aristotle, was born about 372 B.C. at Eresos on the island of Lesbos. He studied at Athens and became an adherent of the school of Plato, and later a friend and pupil of his master Aristotle; when Aristotle withdrew from Athens (before his death in 322), he succeeded him as leader of the Peripatetic school of philosophy. He remained its spokesman and outstanding figure until his death, about 287; for according to Diogenes Laertius,[1] he died at the age of eighty-five.[2]

Though Theophrastus is best known in literature for his *Characters*, a work which has had considerable influence on the drama and on other branches of literature, his writings on natural science are at least of equal importance. His two great works on plants, for example, have led posterity to consider him one of the greatest botanists of all time, the founder of botanical science. Following the practice of the philosophers of his day, he was, however, a voluminous writer on a great variety of subjects. *Primary Propositions, Problems in Natural Philosophy, History of Astronomy, Love, Meteorology, Epilepsy, Animals, Motion, Laws, Odors, Wine and Oil, Proverbs, Water, Fire, History of Geometry, Sleep and Dreams, Virtue, Inventions, Music, Poetry, History of*

[1] *Lives of Eminent Philosophers*, Book V, chap. ii, sec. 40. At the beginning of chap. ii Diogenes Laertius says that he obtained information about Theophrastus from Apollodorus.

[2] The dates 372-287 are accepted by W. von Christ–W. Schmid–O. Stählin, *Geschichte der griechischen Litteratur* in *Handbuch der klassischen Altertumswissenschaft*, 6th ed., Vol. VII, 2, 1 (1920), p. 60. However, the exact dates are not certain. K.O. Brink, *Oxford Classical Dictionary* (1949), p. 896, suggests 372/369-288/285; and O. Regenbogen in Pauly-Wissowa, *Real-Encyclopädie*, Supplemental Vol. VII (1940), p.1357, puts his birth in 372/371 or 371/370 and his death in 288/287 or 287/286.

· 3 ·

Divine Things, Politics, and *Heaven* are only a few of the titles of the 220 works ascribed to him by Diogenes Laertius.[3]

In fact, so varied and enormous was the output of both Theophrastus and Aristotle that many treatises written by their pupils have no doubt been included under their names; but, because of similarities in style and thought, such works can rightly be considered productions of the Peripatetic school and for convenience be ascribed to the masters. The treatise *On Stones* has sometimes been placed in this category, for in style it is more like a set of student's notes than a finished scientific work. Certain of its passages, such as the one in section 68 containing an illustrative story, are obvious abridgments that seem to be mere memoranda written to recall more detailed information. Hence it seems probable that the treatise, as we now have it, is only a set of notes taken down by some student while listening to lectures given by Theophrastus. It is also possible, on the basis of the same internal evidence, that we have before us the personal lecture notes of the master himself. Whether the treatise in its present form is a set of notes taken at lectures or previously written for lectures, it may safely be assumed to represent the actual views of Theophrastus, and he may reasonably be considered the real author.

Internal evidence indicates that the treatise was written near the end of the fourth century B.C., well within the lifetime of Theophrastus. This evidence appears in section 59 in the form of a statement about the time of the discovery by a certain Kallias of a process for refining cinnabar. The validity of this evidence is discussed at length in the notes on that section of the Commentary.

There are comparatively few manuscripts, editions, and translations of the works of Theophrastus, and the treatise *On Stones* has appeared even less often than some of his other works. Only three codices are known to contain it—namely, Vaticanus 1302, Vaticanus 1305, and Vaticanus Urbinas 108—and the versions differ little from each other. Moreover, only two of them are complete, as Vaticanus 1305 ends in the middle of section 43. According to Schneider, these codices were collated by Brandis, who

---

[3] By actual count there are 226 such titles, but six of these are bracketed as probable repetitions by R. D. Hicks, ed., Diogenes Laertius, *Lives of Eminent Philosophers* (Loeb Classical Library, London and New York, 1925), Vol. I, pp. 488-502.

INTRODUCTION

found very little that would improve the text.[4] The date of Vaticanus 1302 is disputed. Devreese and Gianelli put it as early as the twelfth century,[5] but Diels[6] thinks that it is as late as the fourteenth. The other two manuscripts both belong to the fifteenth century. Heinsius claimed that he had made use of a Heidelberg manuscript, but this statement has been disputed.[7]

The first appearance of the treatise *On Stones* in printed form is in the Aldine edition[8] of the works of Aristotle and Theophrastus published at Venice from 1495 to 1498 and reprinted there in 1552. The first Latin translation of the treatise appeared in Paris in 1578; this was the work of Turnebus, who had already published the corresponding Greek text in 1577. This was followed by the edition of Furlanus, published in 1605 at Hanover, containing the Greek text of some of the works of Theophrastus together with a Latin translation and a commentary. And in 1613 a Greek and Latin edition of his works was published by Heinsius at Leyden. This is an unsatisfactory edition which has been severely criticized by both Schneider and Wimmer. Some emendations of the text of the treatise were published by Salmasius (Claude de Saumaise) in 1629 in his *Plinianae Exercitationes*.[9] In 1647 De Laet published at Leyden an annotated Greek and Latin edition of the treatise *On Stones*; this appeared at the beginning of his work *De gemmis et lapidibus libri duo*, which was published as a supplement to the third edition of De Boodt's famous *Gemmarum et lapidum historia*. It is not strictly an independent publication of the treatise.

The first edition in which the text appeared as a single work, as well as the first translation into any modern language and the first extensive commentary, was published by Hill in 1746 at

[4] Schneider, J. G., ed., *Theophrasti Eresii quae supersunt opera* (Leipzig, 1818-1821), Vol. V, p. 146.
[5] We are indebted to the Vatican Library for this information.
[6] See Diels, H., "Aristotelis qui fertur de Melisso Xenophane Gorgia libellus," *Abhandlungen der königlichen Akademie der Wissenschaften zu Berlin, Philosophisch-historische Classe* (1900), p. 5.
[7] Sir Arthur Hort says that "this claim appears to be entirely fictitious." Theophrastus, *Enquiry into Plants* (London and New York, 1916), p. xii.
[8] The full titles of the various editions mentioned here are listed at the end of this Introduction.
[9] A later edition of this work has been consulted, namely, Salmasius, C., *Plinianae Exercitationes in Caii Julii Solini Polyhistora* (Utrecht, 1689).

London. It is still of considerable value for the light it throws on the state of chemical, geological, and mineralogical knowledge in the period during which it was written.[10] Hill's work appeared in a second and final edition in 1774, largely unaltered as regards the text, translation, and commentary, but containing additional matter in the appendix and a fuller index. A French translation of Hill's English version and commentary, but without the Greek text, was published anonymously at Paris in 1754, and a similar German translation with additional notes by Baumgärtner was published at Nürnberg in 1770. Another German translation by Schmieder is said to have been published at Freiberg in 1807,[11] but this seems to be a very rare work, since no other mention of it could be found, and no copy could be located. These few works, which, with the possible exception of the last, are wholly or mostly based upon the labors of Hill, constitute the only past appearance of the treatise *On Stones* as an independent publication.

In the nineteenth century the treatise again appears in the form of a Greek text and Latin translation in Schneider's elaborate edition of the complete works of Theophrastus published at Leipzig in 1818, followed by a fifth volume in 1821. In this fifth volume Schneider was able to make use of some emendations of the text suggested by Adamantios Coraës[12] in his commentary on the *Geography* of Strabo published at Paris in 1819. The Greek text of the treatise also appears in the third volume of Wimmer's Teubner edition published at Leipzig in 1862. In the Didot edition, published at Paris in 1866, this Greek text of Wimmer is reproduced, and Wimmer's parallel Latin translation is given. Of these three important editions, Schneider's is the most valuable for its extensive critical notes and discussions of the readings of earlier editions. In this respect, the two editions of Wimmer are inferior, though the text of Wimmer is somewhat better, and in

---

[10] Its author, John Hill (1716-1775), was a somewhat eccentric litterateur and scientist who was embroiled with many famous men of his time. Though called by his enemies a quack, and by Dr. Johnson a liar, he was, nevertheless, a very learned man and a very able writer. His translation of the treatise *On Stones* brought him to the notice of the Royal Society and won him the friendship of some of its members, which he forfeited by the publication of certain satirical works directed against them.

[11] Schneider, *op. cit.*, Vol. II, p. 578; Vol. IV, p. 535.

[12] Also written Koraës, Koraïs, or Coray.

INTRODUCTION

his Didot edition he gives a Latin translation that is more original than the translation of Schneider, which largely follows the earlier version of Turnebus. In 1902, a French translation of the treatise was published by Mély[13] in a collection of texts and translations of early works on precious stones. In the same year Stephanides published a valuable list of emendations in the Greek periodical *Athena*.[14] These have been added to the list of variant readings in this book. Finally, in 1922, Mieleitner[15] published a German translation, based on Wimmer's text, in an article on the history of mineralogy in ancient and medieval times.

In addition to these complete publications of the treatise in one form or another, excerpts of various parts of it have been published from time to time in various languages in a number of scattered works. The most extensive publication of such excerpts is that of Lenz,[16] who made German translations of many parts of Hill's English translation and added numerous short notes, most of them original, on the significance of the various passages and on the identification of the minerals and localities mentioned by Theophrastus. The most recent is that of Drabkin,[17] who gives an English translation of 17 sections of the treatise. Aside from these partial translations, only a few other studies of parts of the treatise have appeared. Schwarze[18] began a Latin commentary in 1801, and had published seven parts by 1807. In 1896, Stephanides[19] published an important study of the treatise. Ruska's work, *Das Steinbuch des Aristoteles*,[20] deals with a much later treatise on stones incorrectly attributed to Aristotle, but he refers to the treatise of Theophrastus in his introduction. Reference is made in the Commentary to some of these translated excerpts and special studies, particularly to the interpretations advanced by their authors.

[13] Mély, F. de, *Les lapidaires de l'antiquité et du moyen âge* (Paris, 1896-1902), Vol. III, fasc. i.
[14] Stephanides, M. K., *Athena* XIV (1902), 367-71.
[15] Mieleitner, K., *Fortschritte der Mineralogie, Kristallographie und Petrographie* VII (1922), 431-45.
[16] Lenz, H., *Mineralogie der alten Griechen und Römer* (Gotha, 1861), pp. 16-28.
[17] Cohen, M. R., and Drabkin, I. E., *A Source Book of Greek Science* (New York, 1948).
[18] Schwarze, C. A., *De Theophrasti lapidibus commentationes* (Gorlicii, 1801-1807).
[19] Stephanides, M. K., *The Mineralogy of Theophrastus* (in Greek), (Athens, 1896).
[20] Ruska, J., *Das Steinbuch des Aristoteles* (Heidelberg, 1912).

The Greek text which is printed here is almost the same as the one established by Wimmer, but a few minor changes have been made. The three Vatican manuscripts have been collated, as well as the editions of Aldus, Turnebus, Furlanus, Heinsius, De Laet, Hill, and Schneider. All the important variations in the manuscripts and some of the conjectures made by the editors are listed in the critical notes, and the differences between this text and Wimmer's text have also been included. Though the traditional numbering of the sections of the text has been retained, some adjustments have been made in the paragraphs of the translation where the usual divisions between the sections appear to be illogical.

In this translation an attempt has been made to give a clear and simple English version, but at the same time to keep as close as possible to the actual words of the Greek text. In this respect the translation differs from the rather free version of Hill, who reflected the spirit of his times and in many passages preferred elegance of expression to accuracy of statement. Nevertheless, because of the very compressed style of Theophrastus, certain passages in the present translation are of necessity expanded paraphrases of the Greek text. The difficult problem of the translation of the Greek names of mineral substances has been treated in the following way. Names of mineral substances, particularly those of precious stones, for which no exact English equivalent could be found are simply transliterated, and appear in italics in the translation, and the question of their identification is discussed in the Commentary. Names for which an exact English equivalent could be given are so translated and do not appear in italics. Generally the Greek spelling of proper names has been used, but wherever the Latin or the English spelling is customary, this has been preferred; thus Theophrastus and Athens are written instead of *Theophrastos* and *Athenai*.

It has generally been thought that the treatise is a fragment of a very much larger work. A possible explanation of the lacunae, and perhaps of the marked lack of literary style, may be found in an account given by Strabo,[21] who describes the fate of the manuscript books of Theophrastus after his death, and the later un-

---

[21] *Geography*, Book XIII, chap. i, sec. 54.

skillful attempts to restore them to their original state. Apart from a few obvious gaps, however, and the rather abrupt ending, there is no real evidence that the treatise in its present form is not a separate, fairly complete work. Its brevity is apparently the basis of most of the suppositions that the text as we have it is a mere fragment; but if due consideration is given to the nature of the treatise and to the extent of ancient mineralogical knowledge as shown by other sources, it will be seen that it covers the field indicated by its title in an adequate manner, even though it may not be complete.

Without being a purely descriptive or a purely philosophical work, the treatise seems to be an attempt to classify mineral substances on the basis of Aristotelian principles, and a number of specific examples are used, mainly for purposes of illustration, without any intention of giving extended descriptions. It may be inferred that Theophrastus mentions only a small proportion of the mineral products known to him and his contemporaries; for Pliny, though he draws largely from Greek authors, some older than Theophrastus, mentions about ten times as many kinds of rocks or minerals. Those mentioned by Theophrastus appear to be introduced mainly to illustrate in a general way contrasting behavior and distinctive differences in stones and earths, and he may not have intended to catalogue the numerous varieties that were known at the time. This would explain why he describes relatively few mineral substances in any detail, and why he pays so little attention to certain common and highly useful ones about which a good deal must have been known even in his day.

From the historical standpoint the treatise is of special interest because it represents, so far as we know, the first attempt to study mineral substances in a systematic way. For this purpose, Theophrastus divides them into two main classes, stones and earths, discussion of the latter being confined to the second and smaller portion of the treatise. Though few in number, the concise accounts of ancient chemical processes included in this division are of no little importance for the history of chemical technology. At first glance, the structure of the treatise may seem to be loose or even disconnected, but on closer examination it will be readily apparent that this is not so. From the very beginning Theophrastus proceeds in a systematic way to develop the subject under discussion, proceeding regularly from the general to the particular, foreshadowing what is to come and making easy transitions from one phase of the general subject to another. Though his whole method of treatment is logical enough, the classification or system resulting from it, being grounded upon superficial appearance and behavior rather than upon any concept of chemical composition, necessarily has marked limitations. Nevertheless, from a scientific standpoint this little treatise is much better than the other ancient and medieval works on minerals that are known to us. Pliny, for example, though he treats the subject far more extensively, does so in a much less critical and systematic fashion. The comparative freedom of the treatise *On Stones* from fable and magic should be especially noted, for many of the works in this field written centuries later, particularly the medieval lapidaries, dwell largely upon the fancied magical or curative powers of precious stones. In fact, for almost two thousand years this treatise by Theophrastus remained the most rational and systematic attempt at a study of mineral substances.

MANUSCRIPTS AND EDITIONS

MANUSCRIPTS

Codex Vaticanus Graecus 1302, XII-XIV Century.
Codex Vaticanus Graecus 1305, XV Century.
Codex Vaticanus Urbinas Graecus 108, XV Century.

EDITIONS

ALDUS. Editio princeps: Aristoteles et Theophrastus. Venetiis, in domo Aldi Manutii Romani et Graecorum studiosi, 1497, tom. II, fol. 254-60 (et repetita in editione Camotiana apud Aldi filios, 1552, pp. 569-82).[1]

TURNEBUS. Theophrasti De lapidibus. Lutetiae, ex officina Federici Morelli, 1577.
Theophrasti de lapidibus liber, ab Adriano Turnebo latinitate donatus. Lutetiae, ex officina Federici Morelli, 1578.

FURLANUS. Theophrasti Eresii peripateticorum post Aristotelem principis, pleraque antehac Latine numquam, nunc Graece et Latine simul edita. Interpretibus Daniele Furlano Cretensi, Adriano Turnebo. Accesserunt liber De innato spiritu, Aristoteli attributus et Danielis Furlani uberes ad omnia commentarii. Hanoviae, typis Wechelianis, apud Claudium Marnium, 1605.

HEINSIUS. Theophrasti Eresii Graece et Latine opera omnia. Daniel Heinsius textum graecum locis infinitis partim ex ingenio partim e libris emendavit, hiulca supplevit, male concepta recensuit, interpretationem passim interpolavit. Cum indice locupletissimo. Lugduni Batavorum, ex typographio Henrici ab Haestens, impensis Iohannis Orlers, And. Cloucq, et Ioh. Maire, 1613.

DE LAET. Ioannis De Laet Antverpiani De gemmis et lapidibus libri duo quibus praemittitur Theophrasti liber De lapidibus Graece et Latine cum brevibus annotationibus. Lugduni Batavorum, ex officina Ioannis Maire, 1647.

HILL. Theophrastus's History of stones. With an English version and critical and philosophical notes, including the modern history of the gems, etc., described by that author, and of many other of the native fossils. By John Hill. To which are added, two letters: One to Dr. James Parsons, F.R.S. On the colours of the sapphire and turquoise. And the other, to Martin Folkes, Esq., Doctor of Laws, and President of the Royal Society; upon the effects of different menstruums on cop-

[1] There is also an *editio Basiliensis* (J. Oporinus, Basileae, 1541), but this has not been used for the critical notes.

per. Both tending to illustrate the doctrine of gems being coloured by metalline particles. London, printed for C. Davis, 1746.

Second edition, as above, with: . . . Greek index, Observations on the new Swedish acid, and of the stone from which it is obtained, and Idea of a natural and artificial method of fossils, by John Hill, 1774.

SCHNEIDER. Theophrasti Eresii quae supersunt opera et excerpta librorum quatuor tomis comprehensa. Ad fidem librorum editorum et scriptorum emendavit, Historiam et libros VI De causis plantarum coniuncta opera D. H. F. Linkii, excerpta solus, explicare conatus est Io. Gottlob Schneider, Saxo. Lipsiae, sumptibus Frid. Christ. Guil. Vogelii, 1818. Tom. V, 1821.

WIMMER (Teubner). Theophrasti Eresii Opera quae supersunt omnia. Ex recognitione Friderici Wimmer. Tomus tertius fragmenta continens. Lipsiae, sumptibus et typis B. G. Teubneri, 1862.

WIMMER (Didot). Theophrasti Eresii Opera quae supersunt omnia. Graeca recensuit, Latine interpretatus est, indices rerum et verborum absolutissimos adjecit Fridericus Wimmer . . . Parisiis, Firmin Didot, 1866.

FRENCH TRANSLATIONS. Traité des pierres de Théophraste. Traduit du grec; avec des notes physiques et critiques, traduites de l'anglois de M. Hill; auquel on a ajouté deux lettres du même auteur, l'une au docteur Parsons, sur les couleurs du saphir et de la turquoise; et l'autre à M. Folkes, . . . sur les effets des différens menstrues sur le cuivre. Paris, Herissant, 1754.

Les lapidaires de l'antiquité et du moyen âge, F. de Mély, Tome III, fasc. 1, Paris, 1902.

GERMAN TRANSLATIONS. Theophrastus von den Steinen aus dem Griechischen. Nebst Hills physicalischen und critischen Anmerkungen u. einigen in die Naturgeschichte u. Chymie eingeschlagenden Briefen, aus dem englischen übersetzt. Mit Anmerkungen u. einer Abhandlung von der Kunst der Alten in Steinen zu schneiden vermehrt. Von Albert Heinrich Baumgärtner. Nürnberg, 1770.

Karl Mieleitner, "Geschichte der Mineralogie im Altertum und im Mittelalter," *Fortschritte der Mineralogie, Kristallographie und Petrologie* VII (1922), 427-80.

Since the manuscripts correspond in error to a marked degree, they are either derived from a common source or B and C depend on A. A and C are very similar in their readings, but there are certain differences in B. The additional mistakes that are found here are probably due to a careless copyist, but the manuscript

MANUSCRIPTS AND EDITIONS

has also been corrected by a later hand, and sometimes the correct reading appears as follows: μόνος is corrected to ὅμοιος in section 6. The Aldine edition corresponds in error with the manuscripts to such an extent that it is clearly derived from the same source. In a number of places it gives a different reading, frequently the correct one, e.g., χρώμασι for χρώμενα in section 1, but it is not certain whether these changes are due to the skill of the editor or are derived from some other source which has not survived. The most difficult problem occurs in section 20, where Aldus is the sole source for οὐ, which is written as an abbreviation before κισσηροῦται. Wimmer accepts this, though the manuscripts have ἤ or ἥ. There is no evidence that Aldus derived this from another source, and it may be a misprint. He is of no assistance in filling the difficult gaps in the manuscripts, such as the one in section 8 between σχεδόν and λόγον; but though his text contains many misprints, he is frequently helpful in supplying the right reading.

Schneider thinks highly of Turnebus and often accepts his conjectures; Heinsius tends to follow Furlanus, who was not as good an editor as Turnebus. Hill's text is too full of misprints to be reliable.

The variant readings of the three manuscripts and the Aldine edition have been listed wherever they differ from Wimmer's text. Some minor variations in spelling that appear in the manuscripts and some obvious misprints made by Aldus have been omitted. Certain conjectures that appear in the six later editions or in other publications have also been listed, but variations that seem to be due to mistakes or misprints usually have not been included.

Though it is customary to use Latin for critical notes on a text, this has not been done here, since the book is intended for readers interested in science as well as for classical scholars; but for the sake of brevity the abbreviations "om.," "add.," and "conj." have been used for an omission, an addition, and a conjecture. Words added to the text are indicated by pointed brackets, words removed from the text are indicated by square brackets, and doubtful words are marked with a dagger. Since the editors differ in their use of brackets and are sometimes inconsistent, it has not always been easy to guess their meaning.

Wimmer's Didot edition is the same as the Teubner edition except for a few unimportant misprints that have been corrected and two minor variations in section 36 and section 37. A few changes have been made in Wimmer's text, and in these places Wimmer's readings have been listed. Thus ἐξομοιοῦν[ται] has been read in section 4, δἰ ὧν in section 6, ⟨πάντες τῶν κατὰ⟩ in section 8, ἐν τοῖς in section 13, ἐκκαίεται in section 17, ἢ before κισσηροῦται in section 20, πολυτιμοτέρα in section 22, τανῶν in section 25, λευκότατον in section 55, γεωφανέσιν in section 61, and τοιοῦτο in section 65. Sometimes Wimmer omits the definite article where it appears in the manuscripts. It seems better to follow the manuscripts whenever this is possible; therefore τῆς has been restored before κισσήριδος in section 22, τούς before ἐτησίας in section 35, τῷ before ἐλαίῳ in section 42, τῶν before παχυτάτων in section 55, τά before ἀπομάγματα in section 67, and τοῖς before τοιούτοις in section 68. In section 36 Wimmer keeps τούς before πολυτελεῖς in his first edition and omits it in the second. Sometimes the definite article does not appear in the manuscripts, though it is really needed for clarity. In these places ⟨ἡ⟩ has been added to the text as follows: in section 5 before κατὰ τὰς ἐργασίας and in section 22 after μέλαινα, before μαλώδης, and before ἐκ τῆς θαλάσσης. A few accents have been changed, e.g., σπίνος in section 13 and πρασῖτις and αἱματῖτις in section 37.

In the Didot edition Wimmer uses capital letters at the beginning of each sentence, but in the Teubner edition he uses them only at the beginning of chapters. Here the Teubner edition has been followed. Adjectives formed from place names have been written with capitals wherever Wimmer has used small letters, but these changes have not been included in the critical notes.

ABBREVIATIONS

In the critical notes the following abbreviations are used for the manuscripts, and the name of the editor is used for each of the printed editions.

MANUSCRIPTS

1. A = Codex Vaticanus Graecus 1302, 12th-14th century.
2. B = Codex Vaticanus Graecus 1305, 15th century. This ends in the middle of section 43.
3. C = Codex Vaticanus Urbinas Graecus 108, 15th century.
   Ω = the consensus of A, B, C up to section 43 and of A, C from section 43 to section 69.

FIRST EDITION

4. Aldus = Aldus Manutius, Venice, 1497.

OTHER EDITIONS

5. Turnebus = A. Turnebus, Paris. 1577.
6. Furlanus = D. Furlanus, Hanover, 1605.
7. Heinsius = D. Heinsius, Leyden, 1613.
8. Laet = J. de Laet, Leyden, 1647.
9. Hill = J. Hill, second edition, London, 1774.
10. (a) Schneider = J. G. Schneider, Leipzig, 1818.
    (b) Schneider (Syll.) = J. G. Schneider, Syllabus emendandorum et addendorum, Vol. 5, 1821.

EDITIONS USED FOR THE TEXT

11. Wimmer = F. Wimmer, Teubner edition, Leipzig, 1862, and Didot edition, Paris, 1866. Where they differ, they are represented by Wimmer (T) and Wimmer (D).

OTHER AUTHORS AND WORKS

Stephanides = M. K. Stephanides, *Athena*, XIV (Athens, 1902), 367-71.
Stephanides (Min.) = M. K. Stephanides, *The Mineralogy of Theophrastus* (In Greek), Athens, 1896.
Coraës = Adamantios Coraës.
Salmasius = C. Salmasius (Claude de Saumaise).
Eichholz = D. E. Eichholz, *Classical Review*, LXVI (1952), 144-45; LIX (1945), 52.

TEXT

TEXT

ΠΕΡΙ ΛΙΘΩΝ

Τῶν ἐν τῇ γῇ συνισταμένων τὰ μέν ἐστιν ὕδατος, τὰ δὲ γῆς. 1 ὕδατος μὲν τὰ μεταλλευόμενα καθάπερ ἄργυρος καὶ χρυσὸς καὶ τἆλλα, γῆς δὲ λίθος τε καὶ ὅσα λίθων εἴδη περιττότερα, καὶ εἴ τινες δὴ τῆς γῆς αὐτῆς ἰδιώτεραι φύσεις εἰσὶν ἢ χρώμασιν ἢ λειότησιν ἢ πυκνότησιν ἢ ἄλλῃ τινὶ δυνάμει. περὶ μὲν οὖν τῶν μεταλλευομένων ἐν ἄλλοις τεθεώρηται· περὶ δὲ τούτων νῦν λέγωμεν. ἅπαντα οὖν ταῦτα χρὴ νομίζειν ὡς ἁπλῶς εἰπεῖν ἐκ καθαρᾶς 2 τινος συνεστάναι καὶ ὁμαλῆς ὕλης, εἴτε συρροῆς εἴτε διηθήσεώς τινος γενομένης, εἴτε ὡς ἀνωτέρω εἴρηται καὶ κατ' ἄλλον τρόπον ἐκκεκριμένης· τάχα γὰρ ἐνδέχεται τὰ μὲν οὕτως, τὰ δ' ἐκείνως, τὰ δ' ἄλλως. ἀφ' ὧν δὴ καὶ τὸ λεῖον καὶ τὸ πυκνὸν καὶ τὸ στιλπνὸν καὶ διαφανὲς καὶ τἆλλα τὰ τοιαῦτα ἔχουσι, καὶ ὅσῳ ἂν ὁμαλέστερον καὶ καθαρώτερον ἕκαστον ᾖ τοσούτῳ καὶ ταῦτα μᾶλλον ὑπάρχει. τὸ γὰρ ὅλον ὡς ἂν ἀκριβείας ἔχῃ τὰ κατὰ τὴν σύστασιν ἢ πῆξιν οὕτως ἀκολουθεῖ καὶ τὰ ἀπ' ἐκείνων. ἡ δὲ πῆξις τοῖς μὲν 3 ἀπὸ θερμοῦ τοῖς δ' ἀπὸ ψυχροῦ γίνεται. κωλύει γὰρ ἴσως οὐδὲν ἔνια γένη λίθων ὑφ' ἑκατέρων συνίστασθαι τούτων. ἐπεὶ τά γε τῆς γῆς ἅπαντα δόξειεν ἂν ὑπὸ πυρός· ἐπείπερ τοῖς ἐναντίοις ἑκάστων ἡ πῆξις καὶ ἡ τῆξις. ἰδιότητες δὲ πλείους εἰσὶν ἐν τοῖς λίθοις· ἐν μὲν γὰρ τῇ γῇ χρώμασί τε καὶ γλισχρότητι καὶ λειότητι καὶ πυκνότητι καὶ τοῖς τοιούτοις αἱ πολλαὶ διαφοραί, κατὰ δὲ τὰ ἄλλα σπάνιοι. τοῖς δὲ λίθοις αὗταί τε καὶ πρὸς ταύταις αἱ 4 κατὰ τὰς δυνάμεις τοῦ τε ποιεῖν ἢ πάσχειν ἢ τοῦ μὴ πάσχειν. τηκτοὶ γὰρ οἱ δ' ἄτηκτοι, καὶ καυστοὶ οἱ δ' ἄκαυστοι, καὶ ἄλλα τούτοις ὅμοια. καὶ ἐν αὐτῇ τῇ καύσει καὶ πυρώσει πλείους ἔχοντες διαφοράς, ἔνιοι δὲ τοῖς χρώμασιν ἐξομοιοῦν[ται] δυνάμενοι τὸ ὕδωρ ὥσπερ ἡ σμάραγδος, οἱ δ' ὅλως ἀπολιθοῦν τὰ τιθέμενα εἰς ἑαυτούς, ἕτεροι δὲ ὁλκήν τινα ποιεῖν, οἱ δὲ βασανίζειν τὸν χρυσὸν καὶ τὸν ἄργυρον ὥσπερ ἥ τε καλουμένη λίθος Ἡράκλεια καὶ ἡ Λυδή. θαυμασιωτάτη δὲ καὶ μεγίστη δύναμις, εἴπερ ἀληθές, 5 ἡ τῶν τικτόντων· γνωριμωτέρα δὲ [τῶν] καὶ ἐν πλείοσι ⟨ἡ⟩ κατὰ τὰς ἐργασίας· γλυπτοὶ γὰρ ἔνιοι καὶ τορνευτοὶ καὶ πριστοί, τῶν δὲ οὐδὲ ὅλως ἅπτεται σιδήριον, ἐνίων δὲ κακῶς καὶ μόλις. εἰσὶ δὲ πλείους καὶ ἄλλαι παρὰ ταύτας διαφοραί. αἱ μὲν οὖν κατὰ 6

τὰ χρώματα καὶ τὰς σκληρότητας καὶ μαλακότητας καὶ λειότητας καὶ τἆλλα τὰ τοιαῦτα, δι' ὧν τὸ περιττόν, πλείοσιν ὑπάρχουσι καὶ ἐνίοις γε κατὰ τόπον ὅλον. ἐξ ὧν δὴ καὶ διωνομασμέναι λιθοτομίαι Παρίων τε καὶ Πεντελικῶν καὶ Χίων τε καὶ Θηβαϊκῶν, καὶ ὡς ὁ περὶ Αἴγυπτον ἐν Θήβαις ἀλαβαστρίτης,—καὶ γὰρ οὗτος μέγας τέμνεται,—καὶ ὁ τῷ ἐλέφαντι ὅμοιος ὁ χερνίτης καλού-
7 μενος, ἐν ᾗ πυέλῳ φασὶ καὶ Δαρεῖον κεῖσθαι. καὶ ὁ πόρος ὅμοιος τῷ χρώματι καὶ τῇ πυκνότητι τῷ Παρίῳ τὴν δὲ κουφότητα μόνον ἔχων τοῦ πόρου, διὸ καὶ ἐν τοῖς σπουδαζομένοις οἰκήμασιν ὥσπερ διάζωμα τιθέασιν αὐτὸν οἱ Αἰγύπτιοι. καὶ μέλας αὐτόθι διαφανὴς ὅμοιος τῷ Χίῳ, καὶ παρ' ἄλλοις δὲ ἕτεροι πλείους. αἱ μὲν οὖν τοιαῦται διαφοραὶ καθάπερ ἐλέχθη κοινότεραι πλείοσιν, αἱ δὲ κατὰ τὰς δυνάμεις τὰς προειρημένας οὐκέτι τόποις ὅλοις ὑπάρχουσιν
8 οὐδὲ συνεχείαις λίθων οὐδὲ μεγέθεσιν. ἔνιοι δὲ καὶ σπάνιοι πάμπαν εἰσὶ καὶ σμικροὶ καθάπερ ἥ τε σμάραγδος καὶ τὸ σάρδιον καὶ ὁ ἄνθραξ καὶ ἡ σάπφειρος καὶ σχεδὸν ⟨πάντες τῶν κατὰ⟩ λόγον εἰς τὰ σφραγίδια γλυπτῶν. οἱ δὲ καὶ ἐν ἑτέροις εὑρίσκονται διακοπτομένοις. ὀλίγοι δὲ καὶ οἱ περὶ τὴν πύρωσιν καὶ καῦσιν, ὑπὲρ ὧν δὴ καὶ πρῶτον ἴσως λεκτέον τίνας καὶ πόσας ἔχουσι διαφοράς.
9 Κατὰ δὴ τὴν πύρωσιν οἱ μὲν τήκονται καὶ ῥέουσιν ὥσπερ οἱ μεταλλευτοί. ῥεῖ γὰρ ἅμα τῷ ἀργύρῳ καὶ τῷ χαλκῷ καὶ σιδήρῳ καὶ ἡ λίθος ἡ ἐκ τούτων, εἴτ' οὖν διὰ τὴν ὑγρότητα τῶν ἐνυπαρχόντων εἴτε καὶ δι' αὑτούς. ὡσαύτως δὲ καὶ οἱ πυρομάχοι καὶ οἱ μυλίαι ῥέουσιν οἷς ἐπιτιθέασιν οἱ καίοντες. οἱ δὲ καὶ ὅλως λέγουσι πάντας τήκεσθαι πλὴν τοῦ μαρμάρου, τοῦτον δὲ κατακαίεσθαι
10 καὶ κονίαν ἐξ αὐτοῦ γίνεσθαι. δόξειε δ' ἂν οὕτως ἐπὶ πλεῖον εἰρῆσθαι· πολλοὶ γὰρ οἱ ῥηγνύμενοι καὶ διαπηδῶντες ὡς ἀπομαχόμενοι τὴν πύρωσιν ὥσπερ [οὐδ'] ὁ κέραμος. ὃ καὶ κατὰ λόγον ἐστίν, οἵτινες ἐξυγρασμένοι τυγχάνουσιν· τὸ γὰρ τηκτὸν ἔνικμον εἶναι
11 δεῖ καὶ ὑγρότητ' ἔχειν πλείω. φασὶ δὲ καὶ τῶν ἡλιουμένων τοὺς μὲν ἀναξηραίνεσθαι τελείως ὥστ' ἀχρείους εἶναι μὴ καταβρεχθέντας πάλιν καὶ συνικμασθέντας τοὺς δὲ καὶ μαλακωτέρους καὶ διαθραύστους μᾶλλον. φανερὸν δὲ ὡς ἀμφοτέρων μὲν ἐξαιρεῖται τὴν ὑγρότητα, συμβαίνει δὲ τοὺς μὲν πυκνοὺς ἀποξηραινομένους σκληρύνεσθαι, τοὺς δὲ μανοὺς καὶ ὧν ἡ σύμφυσις τοιαύτη θραυσ-
12 τοὺς εἶναι καὶ τηκτούς. ἔνιοι δὲ τῶν θραυστῶν ἀνθρακοῦνται τῇ καύσει καὶ διαμένουσι πλείω χρόνον ὥσπερ οἱ περὶ Βίνας ἐν τῷ

μετάλλῳ οὓς ὁ ποταμὸς καταφέρει· καίονται γὰρ ὅταν ἄνθρακες ἐπιτεθῶσι καὶ μέχρι τούτου ἄχρις ἂν φυσᾷ τις, εἶτ' ἀπομαραίνονται καὶ πάλιν καίονται, διὸ καὶ πολὺν χρόνον ἡ χρῆσις· ἡ δ' ὀσμὴ βαρεῖα σφόδρα καὶ δυσχερής. ὃν δὲ καλοῦσι σπίνον, 13 ὃς ἦν ἐν τοῖς μετάλλοις, οὗτος διακοπεὶς καὶ συντεθεὶς πρὸς ἑαυτὸν ἐν τῷ ἡλίῳ τιθέμενος καίεται, καὶ μᾶλλον ἐὰν ἐπιψεκάσῃ καὶ περιράνῃ τις. ὁ δὲ Λιπαραῖος ἐκφοροῦταί τε τῇ καύσει καὶ 14 γίνεται κισσηροειδὴς ὥσθ' ἅμα τήν τε χρόαν μεταβάλλειν καὶ τὴν πυκνότητα· μέλας τε γὰρ καὶ λεῖός ἐστι καὶ πυκνὸς ἄκαυστος ὤν. γίνεται δὲ οὗτος ἐν τῇ κισσήρει διειλημμένος ἄλλοθι καὶ ἄλλοθι καθάπερ ἐν κυττάρῳ καὶ οὐ συνεχής, ὥσπερ καὶ ἐν Μήλῳ φασὶ τὴν κίσσηριν ἐν ἄλλῳ τινὶ λίθῳ γίνεσθαι. καὶ ἐκεῖνος μὲν τούτῳ ὥσπερ ἀντιπεπονθώς· πλὴν ὁ λίθος οὗτος οὐχ ὅμοιος τῷ Λιπαραίῳ. ἐκφοροῦται δὲ καὶ ὁ ἐν Τετράδι τῆς Σικελίας 15 γινόμενος· τοῦτο δὲ τὸ χωρίον ἐστὶ κατὰ Λιπάραν, ὁ δὲ λίθος ἐν τῇ ἄκρᾳ τῇ Ἐρινεάδι καλουμένῃ πολὺς ὁμοίως τῷ ἐν Βίναις καιόμενος ὀσμὴν ἀφίησιν ἀσφάλτου, τὸ δ' ἐκ τῆς κατακαύσεως ὅμοιον γίνεται γῇ κεκαυμένῃ. οὓς δὲ καλοῦσιν εὐθὺς ἄνθρακας 16 τῶν ὀρυττομένων διὰ τὴν χρείαν εἰσὶ γεώδεις, ἐκκαίονται δὲ καὶ πυροῦνται καθάπερ οἱ ἄνθρακες. εἰσὶ δὲ περί τε τὴν Λιγυστικὴν ὅπου καὶ τὸ ἤλεκτρον, καὶ ἐν τῇ Ἠλείᾳ βαδιζόντων Ὀλυμπίαζε τὴν δι' ὄρους, οἷς καὶ οἱ χαλκεῖς χρῶνται. εὑρέθη δέ ποτε ἐν 17 τοῖς Σκαπτησύλης μετάλλοις λίθος ὃς τῇ μὲν ὄψει παρόμοιος ὢν ξύλῳ σαπρῷ, ὅτε δ' ἐπιχέοιτό τις ἔλαιον ἐκκαίεται, καὶ ὅτ' ἐκκαυθείη τότε παύεται καὶ αὐτὸς ὥσπερ ἀπαθὴς ὤν. τῶν μὲν οὖν καιομένων σχεδὸν αὗται διαφοραί.

Ἄλλο δέ τι γένος ἐστὶ λίθων ὥσπερ ἐξ ἐναντίων πεφυκός, 18 ἄκαυστον ὅλως, ἄνθραξ καλούμενος, ἐξ οὗ καὶ τὰ σφραγίδια γλύφουσιν, ἐρυθρὸν μὲν τῷ χρώματι, πρὸς δὲ τὸν ἥλιον τιθέμενον ἄνθρακος καιομένου ποιεῖ χρόαν. τιμιώτατον δ' ὡς εἰπεῖν· μικρὸν γὰρ σφόδρα τετταράκοντα χρυσῶν. ἄγεται δὲ οὗτος ἐκ Καρχηδόνος καὶ Μασσαλίας. οὐ καίεται δὲ ὁ περὶ Μίλητον γωνιοειδὴς 19 ὢν ἐν ᾧπερ καὶ τὰ ἐξάγωνα. καλοῦσι δ' ἄνθρακα καὶ τοῦτον, ὃ καὶ θαυμαστόν ἐστιν· ὅμοιον γὰρ τρόπον τινὰ καὶ τὸ τοῦ ἀδάμαντος· οὐ γὰρ οὐδ' ὥσπερ ἡ κίσσηρις καὶ τέφρα δόξειεν ἂν διὰ τὸ μηδὲν ἔχειν ὑγρόν· ταῦτα γὰρ ἄκαυστα καὶ ἀπύρωτα διὰ τὸ ἐξῃρῆσθαι τὸ ὑγρόν· ἐπεὶ καὶ τὸ ὅλον ἡ κίσσηρις ἐκ κατακαύσεως δοκεῖ τισι γίνεσθαι, πλὴν τῆς ἐκ τοῦ ἀφροῦ τῆς θαλάσ-

20 σης συνισταμένης. λαμβάνουσι δὲ τὴν πίστιν διὰ τῆς αἰσθήσεως ἔκ τε τῶν περὶ τοὺς κρατῆρας γινομένων καὶ ἐκ τῆς †διαβάρου λίθου τῆς φλογουμένης† ἢ κισσηροῦται. μαρτυρεῖν δὲ καὶ οἱ τόποι δοκοῦσιν ἐν οἷς ἡ γένεσις· καὶ γὰρ ἐν τοῖς † . . .
21 μάλιστα καὶ ἡ κίσσηρις. τάχα δὲ ἡ μὲν οὕτως αἱ δ' ἄλλως καὶ πλείους τρόποι τῆς γενέσεως. ἡ γὰρ ἐν Νισύρῳ καθάπερ ἐξ ἄμμου τινὸς ἔοικε συγκεῖσθαι. σημεῖον δὲ λαμβάνουσιν ὅτι τῶν εὑρισκομένων ἔνιαι διαθρύπτονται ἐν ταῖς χερσὶν ὥσπερ εἰς ἄμμον διὰ τὸ μήπω συνεστάναι μηδὲ συμπεπηγέναι. εὑρίσκουσι δ' ἀθρόας κατὰ μικρὰ χειροπληθεῖς ὅσον πολλὰς ἢ μικρῷ μείζους ὅταν ἀπαμήσωνται τἄνω· ἐλαφρὰ δὲ σφόδρα ἡ ἄμμος. ἡ δ' αὖ καὶ ἐν Μήλῳ πᾶσα μὲν † . . . ἔνια δ' αὖ ἐν λίθῳ τινὶ ἑτέρῳ
22 γίνεται καθάπερ ἐλέχθη πρότερον. διαφορὰς δ' ἔχουσι πρὸς ἀλλήλας καὶ χρώματι καὶ πυκνότητι καὶ βάρει· χρώματι μὲν ὅτι μέλαινα ⟨ἡ⟩ ἐκ τοῦ ῥύακος τοῦ ἐν Σικελίᾳ· πυκνότητι δὲ καὶ βάρει αὕτη τε καὶ ⟨ἡ⟩ †μαλώδης. γίνεται γὰρ τις καὶ τοιαύτη κίσσηρις καὶ βάρος ἔχει καὶ πυκνότητα καὶ ἐν τῇ χρήσει πολυτιμοτέρα τῆς ἑτέρας. τμητικὴ δὲ καὶ ἡ ἐκ τοῦ ῥύακος μᾶλλον τῆς κούφης καὶ λευκῆς, τμητικωτάτη δ' ⟨ἡ⟩ ἐκ τῆς θαλάσσης αὐτῆς. καὶ περὶ μὲν τῆς κισσήριδος ἐπὶ τοσοῦτον εἰρήσθω. περὶ δὲ τῶν πυρουμένων καὶ τῶν ἀπυρώτων λίθων ἀφ' ὧν καὶ εἰς τοῦτο ἐξέβημεν ἐν ἄλλοις θεωρητέον τὰς αἰτίας.
23 Τῶν δὲ λίθων καὶ ἄλλαι ⟨διάφοροι⟩ τυγχάνουσιν ἐξ ὧν καὶ τὰ σφραγίδια γλύφουσιν. αἱ μὲν τῇ ὄψει μόνον οἷον τὸ σάρδιον καὶ ἡ ἴασπις καὶ ἡ σάπφειρος· αὕτη δ' ἐστὶν ὥσπερ χρυσόπαστος. ἡ δὲ σμάραγδος καὶ δυνάμεις τινὰς ἔχει· τοῦ τε γὰρ ὕδατος ὥσπερ εἴπομεν ἐξομοιοῦται τὴν χρόαν ἑαυτῇ, μετρία μὲν οὖσα ἐλάττονος, ἡ δὲ μεγίστη παντός, ἡ δὲ χειρίστη τοῦ καθ' αὑτὴν
24 μόνον. καὶ πρὸς τὰ ὄμματα ἀγαθή, διὸ καὶ τὰ σφραγίδια φοροῦσιν ἐξ αὐτῆς ὥστε βλέπειν· ἔστι δὲ σπανία καὶ τὸ μέγεθος οὐ μεγάλη, πλὴν εἰ πιστεύειν ταῖς ἀναγραφαῖς δεῖ ὑπὲρ τῶν βασιλέων τῶν Αἰγυπτίων· ἐκείνοις γάρ φασι κομισθῆναί ποτ' ἐν δώροις παρὰ τοῦ Βαβυλωνίων βασιλέως μῆκος μὲν τετράπηχυν πλάτος δὲ τρίπηχυν. ἀνακεῖσθαι δὲ καὶ ἐν τῷ τοῦ Διὸς ὀβελίσκῳ σμαράγδους τέτταρας, μῆκος μὲν τετταράκοντα πηχῶν, εὖρος δὲ τῇ μὲν τέτταρας τῇ δὲ δύο. ταῦτα μὲν οὖν ὅτι κατὰ τὴν
25 ἐκείνων γραφήν. τῶν δὲ †τανῶν καλουμένων ὑπὸ πολλῶν ἡ ἐν Τύρῳ μεγίστη. στήλη γάρ ἐστιν εὐμεγέθης ἐν τῷ τοῦ Ἡρα-

κλέους ἱερῷ· εἰ μὴ ἄρα ψευδὴς σμάραγδος, καὶ γὰρ τοιαύτη γίνεταί τις φύσις. γίνεται δὲ ἐν τοῖς ἐν ἐφικτῷ καὶ γνωρίμοις τόποις διτταχοῦ μάλιστα περί τε Κύπρον ἐν τοῖς χαλκορυχείοις καὶ ἐν τῇ νήσῳ τῇ ἐπικειμένῃ Χαλκηδόνι. καὶ ἰδιωτέρους εὑρίσκουσιν ἐν ταύτῃ· μεταλλεύεται γὰρ ὥσπερ τἆλλα καὶ ἡ φύσις κατὰ ῥάβδους ἐποίησεν ἐν Κύπρῳ αὐτὴν καθ' αὑτὴν πολλάς. εὑρίσκονται δὲ σπάνιαι μέγεθος ἔχουσαι σφραγίδος ἀλλ' ἐλάττους αἱ πολλαί, διὸ καὶ πρὸς τὴν κόλλησιν αὐτῇ χρῶνται τοῦ χρυσίου· κολλᾷ γὰρ ὥσπερ ἡ χρυσοκόλλα. καὶ ἔνιοί γε δὴ καὶ ὑπολαμβάνουσι τὴν αὐτὴν φύσιν εἶναι· καὶ γὰρ τὴν χρόαν παρόμοιαι τυγχάνουσιν. ἀλλ' ἡ μὲν χρυσοκόλλα δαψιλὴς καὶ ἐν τοῖς χρυσείοις καὶ ἔτι μᾶλλον ἐν τοῖς χαλκορυχείοις ὥσπερ ἐν τοῖς περὶ τοὺς † ... τόπους. ἡ δὲ σμάραγδος σπανία καθάπερ εἴρηται· δοκεῖ γὰρ ἐκ τῆς ἰάσπιδος γίνεσθαι. φασὶ γὰρ εὑρεθῆναί ποτε ἐν Κύπρῳ λίθον ἧς τὸ μὲν ἥμισυ σμάραγδος ἦν τὸ ἥμισυ δὲ ἴασπις ὡς οὔπω μεταβεβληκυίας ἀπὸ τοῦ ὕδατος. ἔστι δέ τις αὐτῆς ἐργασία πρὸς τὸ λαμπρόν· ἀργὴ γὰρ οὖσα οὐ λαμπρά.

Αὕτη τε δὴ περιττὴ τῇ δυνάμει καὶ τὸ λυγγούριον· καὶ γὰρ ἐκ τούτου γλύφεται τὰ σφραγίδια καὶ ἔστι στερεωτάτη καθάπερ λίθος· ἕλκει γὰρ ὥσπερ τὸ ἤλεκτρον, οἱ δέ φασιν οὐ μόνον κάρφη καὶ ξύλον ἀλλὰ καὶ χαλκὸν καὶ σίδηρον ἐὰν ᾖ λεπτός, ὥσπερ καὶ Διοκλῆς ἔλεγεν. ἔστι δὲ διαφανές τε σφόδρα καὶ ψυχρόν. βέλτιον δὲ τὸ τῶν ἀγρίων ἢ τὸ τῶν ἡμέρων καὶ τὸ τῶν ἀρρένων ἢ τὸ τῶν θηλειῶν ὡς καὶ τῆς τροφῆς διαφερούσης, καὶ τοῦ πονεῖν ἢ μὴ πονεῖν, καὶ τῆς τοῦ σώματος ὅλως φύσεως, ᾗ τὸ μὲν ξηρότερον τὸ δ' ὑγρότερον. εὑρίσκουσι δ' ἀνορύττοντες οἱ ἔμπειροι· κατακρύπτεται γὰρ καὶ ἐπαμᾶται γῆν ὅταν οὐρήσῃ. γίνεται δὲ καὶ κατεργασία τις αὐτοῦ πλείων. ἐπεὶ δὲ καὶ τὸ ἤλεκτρον λίθος, τὸ γὰρ ὀρυκτὸν ὃ περὶ Λιγυστικήν, καὶ τούτῳ ἂν ἡ τοῦ ἕλκειν δύναμις ἀκολουθοίη. μάλιστα δ' ἐπίδηλος καὶ φανερωτάτη ἡ τὸν σίδηρον ἄγουσα. γίνεται δὲ καὶ αὕτη σπανία καὶ ὀλιγαχοῦ. καὶ αὕτη μὲν δὴ συναριθμείσθω τὴν δύναμιν ὁμοίαν ἔχειν. ἐξ ὧν δὲ τὰ σφραγίδια ποιεῖται καὶ ἄλλαι πλείους εἰσίν, οἷον ἥ θ' ὑαλοειδὴς ἢ καὶ ἔμφασιν ποιεῖ καὶ διάφασιν, καὶ τὸ ἀνθράκιον, καὶ ἡ ὄμφαξ· ἔτι δὲ καὶ ἡ κρύσταλλος καὶ τὸ ἀμέθυσον, ἄμφω δὲ διαφανῆ, εὑρίσκονται δὲ καὶ αὗται καὶ τὸ σάρδιον διακοπτομένων τινῶν πετρῶν. καὶ ἄλλαι δὲ ὡς προείρηται πρό-

τερον διαφορὰς ἔχουσαι καὶ συνώνυμοι πρὸς ἀλλήλας. τοῦ γὰρ σαρδίου τὸ μὲν διαφανὲς ἐρυθρότερον δὲ καλεῖται θῆλυ, τὸ δὲ
31 διαφανὲς μὲν μελάντερον δὲ [καὶ] ἄρσεν. καὶ τὰ λυγγούρια δὲ ὡσαύτως ὧν τὸ θῆλυ διαφανέστερον καὶ ξανθότερον. καλεῖται δὲ καὶ κύανος ὁ μὲν ἄρρην ὁ δὲ θῆλυς· μελάντερος δὲ ὁ ἄρρην. τὸ δ' ὀνύχιον μικτὸν λευκῷ καὶ φαιῷ παρ' ἄλληλα. τὸ δ' ἀμέθυσον οἰνωπὸν τῇ χρόᾳ. καλὸς δὲ λίθος καὶ ὁ ἀχάτης ὁ ἀπὸ τοῦ Ἀχάτου
32 ποταμοῦ τοῦ ἐν Σικελίᾳ καὶ πωλεῖται τίμιος. ἐν Λαμψάκῳ δέ ποτ' ἐν τοῖς χρυσίοις εὑρέθη θαυμαστὴ λίθος ἐξ ἧς ἀνενεχθείσης πρὸς †στιρὰν σφραγίδιον γλυφθὲν ἀνεπέμφθη βασιλεῖ διὰ τὸ περιττόν.
33 Καὶ αὗται μὲν ἅμα τῷ καλῷ καὶ τὸ σπάνιον ἔχουσιν. αἱ δὲ δὴ ἐκ τῆς Ἑλλάδος εὐτελέστεραι, οἷον τὸ ἀνθράκιον τὸ ἐξ Ὀρχομενοῦ τῆς Ἀρκαδίας. ἔστι δὲ οὗτος μελάντερος τοῦ Χίου· κάτοπτρα δὲ ἐξ αὐτοῦ ποιοῦσι· καὶ ὁ Τροιζήνιος· οὗτος δὲ ποικίλος τὰ μὲν φοινικοῖς τὰ δὲ λευκοῖς χρώμασι. ποικίλος δὲ καὶ ὁ Κορίν-
34 θιος τοῖς αὐτοῖς χρώμασι πλὴν ὅτι χλωροειδέστερος. τὸ δὲ ὅλον πολλοὶ τυγχάνουσιν οἱ τοιοῦτοι ἀλλ' οἱ περιττοὶ σπάνιοι καὶ ἐξ ὀλίγων τόπων οἷον ἐκ τε Καρχηδόνος καὶ ἐκ τῶν περὶ Μασσαλίαν καὶ ἐξ Αἰγύπτου κατὰ τοὺς Καταδούπους καὶ Συήνης πρὸς
35 Ἐλεφαντίνῃ πόλει καὶ ἐκ τῆς Ψεφὼ καλουμένης χώρας. καὶ ἐν Κύπρῳ ἥ τε σμάραγδος καὶ ἡ ἴασπις. οἷς δὲ εἰς τὰ λιθοκόλλητα χρῶνται ἐκ τῆς Βακτριανῆς εἰσὶ πρὸς τῇ ἐρήμῳ. συλλέγουσι δὲ αὐτοὺς ὑπὸ τοὺς ἐτησίας ἱππεῖς ἐξιόντες· τότε γὰρ ἐμφανεῖς γίνονται κινουμένης τῆς ἄμμου διὰ τὸ μέγεθος τῶν πνευμάτων.
36 εἰσὶ δὲ μικροὶ καὶ οὐ μεγάλοι. τῶν σπουδαζομένων δὲ λίθων ἐστὶ καὶ ὁ μαργαρίτης καλούμενος, διαφανὴς μὲν τῇ φύσει, ποιοῦσι δ' ἐξ αὐτοῦ τοὺς πολυτελεῖς ὅρμους. γίνεται δὲ ἐν ὀστρείῳ τινὶ παραπλησίῳ ταῖς πίνναις ⟨πλὴν ἐλάττονι· μέγεθος δὲ ἡλίκος ἰχθύος ὀφθαλμὸς εὐμεγέθης⟩, φέρει δὲ ἥ τε Ἰνδικὴ χώρα καὶ νῆσοί τινες τῶν ἐν τῇ Ἐρυθρᾷ. τὸ μὲν οὖν περιττὸν σχεδὸν ἐν
37 ταύταις. εἰσὶ δὲ καὶ ἄλλαι τινές, οἷον ὁ ἐλέφας ὁ ὀρυκτὸς ποικίλος μέλανι καὶ λευκῷ. καὶ ἣν καλοῦσι σάπφειρον· αὕτη γὰρ μέλαινα οὐκ ἄγαν πόρρω τοῦ κυάνου τοῦ ἄρρενος καὶ πρασῖτις· αὕτη δὲ ἰώδης τῇ χρόᾳ. πυκνὴ δὲ καὶ αἱματῖτις· αὕτη δ' αὐχμώδης καὶ κατὰ τοὔνομα ὡς αἵματος ξηροῦ πεπηγότος. ἄλλη δὲ ἡ καλουμένη ξανθή, οὐ ξανθὴ μὲν τὴν χρόαν, ἔκλευκος δὲ μᾶλλον,
38 ὁ καλοῦσι χρῶμα οἱ Δωριεῖς ξανθόν. τὸ γὰρ κουράλιον, καὶ γὰρ

τοῦθ' ὥσπερ λίθος, τῇ χρόᾳ μὲν ἐρυθρόν, περιφερὲς δ' ὡς ῥίζα· φύεται δ' ἐν τῇ θαλάττῃ. τρόπον δέ τινα οὐ πόρρω τούτου τῇ φύσει καὶ ὁ Ἰνδικὸς κάλαμος ἀπολελιθωμένος. ταῦτα μὲν οὖν ἄλλης σκέψεως.

Τῶν δὲ λίθων πολλαί τινες αἱ φύσεις καὶ τῶν μεταλλευομένων. 39 ἔνιαι γὰρ ἅμα χρυσὸν ἔχουσι καὶ ἄργυρον, προφανῆ δὲ μόνον ἄργυρον· βαρύτεραι δὲ αὗται πολὺ καὶ τῇ ῥοπῇ καὶ τῇ ὀσμῇ· καὶ κύανος αὐτοφυὴς ἔχων ἐν ἑαυτῷ χρυσοκόλλαν. ἄλλη δὲ λίθος ὁμοία τὴν χρόαν τοῖς ἄνθραξι· βάρος δὲ ἔχουσι. τὸ δὲ ὅλον ἐν 40 τοῖς μετάλλοις πλεῖσται καὶ ἰδιώταται φύσεις εὑρίσκονται τῶν τοιούτων, ὧν τὰ μέν εἰσι γῆς καθάπερ ὤχρα καὶ μίλτος, τὰ δὲ οἷον ἄμμου καθάπερ χρυσοκόλλα καὶ κύανος, τὰ δὲ κονίας οἷον σανδαράκη καὶ ἀρρενικὸν καὶ ὅσα ὅμοια τούτοις. καὶ τῶν μὲν τοιούτων πλείους ἄν τις λάβοι τὰς ἰδιότητας. ἔνιαι δὲ λίθοι καὶ 41 τὰς τοιαύτας ἔχουσι δυνάμεις εἰς τὸ μὴ πάσχειν, ὥσπερ εἴπομεν, οἷον τὸ μὴ γλύφεσθαι σιδηρίοις ἀλλὰ λίθοις ἑτέροις. ὅλως μὲν ἡ κατὰ τὰς ἐργασίας καὶ τῶν μειζόνων λίθων πολλὴ διαφορά. πριστοὶ γάρ, οἱ δὲ γλυπτοί, καθάπερ ἐλέχθη, καὶ τορνευτοὶ τυγχάνουσι, καθάπερ καὶ ἡ Μαγνῆτις αὕτη λίθος ἡ καὶ ὄψει περιττὸν ἔχουσα, καὶ ἧς γε δή τινες θαυμάζουσι τὴν ὁμοίωσιν τῷ ἀργύρῳ μηδαμῶς οὔσης συγγενοῦς. πλείους δ' εἰσὶν οἱ δεχόμενοι πάσας 42 τὰς ἐργασίας. ἐπεὶ καὶ ἐν Σίφνῳ τοιοῦτός τίς ἐστιν ὀρυκτὸς ὡς τρία στάδια ἀπὸ θαλάττης, στρογγύλος καὶ βωλώδης, καὶ τορνεύεται καὶ γλύφεται διὰ τὸ μαλακόν· ὅταν δὲ πυρωθῇ καὶ ἀποβαφῇ τῷ ἐλαίῳ, μέλας τε σφόδρα γίνεται καὶ σκληρός. ποιοῦσι δ' ἐξ αὐτοῦ σκεύη τὰ ἐπιτράπεζα. οἱ μὲν ⟨οὖν⟩ τοιοῦτοι πάντες 43 προσδέχονται τὴν τοῦ σιδήρου δύναμιν· ἔνιοι δὲ λίθοις ἄλλοις γλύφονται, σιδήροις δ' οὐ δύνανται καθάπερ εἴπομεν. οἱ δὲ σιδήροις μὲν ἀμβλυτέροις δέ· καὶ εἰσὶν † ... παραπλησίως δὲ †κάτω τὰ ... μὴ τέμνεσθαι ... σιδήρῳ· καίτοι καὶ †στερεὸν ἔτε ... ἰσχυρότερα τέμνει καὶ σίδηρος λίθου σκληρότερος ὤν. ἄτοπον δὲ κἀκεῖνο φαίνεται διότι ἡ μὲν ἀκόνη κατεσθίει τὸν 44 σίδηρον, ὁ δὲ σίδηρος ταύτην μὲν δύναται διαιρεῖν καὶ ῥυθμίζειν, ἐξ ἧς δὲ αἱ σφραγῖδες οὔ. καὶ πάλιν ὁ λίθος ᾧ γλύφουσι τὰς σφραγῖδας ἐκ τούτου ἐστὶν ἐξ οὗπερ αἱ ἀκόναι, ἢ ἐξ ὁμοίου τούτῳ· ἄγεται δὲ ἡ ⟨ἀρίστη⟩ ἐξ Ἀρμενίας. θαυμαστὴ δὲ φύσις 45 καὶ τῆς βασανιζούσης τὸν χρυσόν· δοκεῖ γὰρ δὴ τὴν αὐτὴν ἔχειν τῷ πυρὶ δύναμιν· καὶ γὰρ ἐκεῖνο δοκιμάζει. διὸ καὶ ἀπο-

ρουσί τινες οὐκ ἄγαν οἰκείως ἀποροῦντες. οὐ γὰρ τὸν αὐτὸν τρόπον δοκιμάζει, ἀλλὰ τὸ μὲν πῦρ τῷ τὰ χρώματα μεταβάλλειν καὶ ἀλλοιοῦν, ὁ δὲ λίθος τῇ παρατρίψει· δύναται γὰρ ὡς ἔοικεν
46 ἐκλαμβάνειν τὴν ἑκάστου φύσιν. εὑρῆσθαι δέ φασι νῦν ἀμείνω πολὺ τῆς πρότερον ὥστε μὴ μόνον τὸν ἐκ τῆς καθάρσεως ἀλλὰ καὶ τὸν κατάχαλκον χρυσὸν καὶ ἄργυρον γνωρίζειν καὶ πόσον εἰς τὸν στατῆρα μέμικται. σημεῖα δ' ἐστὶν αὐτοῖς ἀπὸ τοῦ ἐλαχίστου· ἐλάχιστον δὲ γίνεται κριθή, εἶτα κόλλυβος, εἶτα τεταρ-
47 τημόριον ἢ ἡμιώβολος, ἐξ ὧν γνωρίζουσι τὸ καθῆκον. εὑρίσκονται δὲ τοιαῦται πᾶσαι ἐν τῷ ποταμῷ Τμώλῳ. λεία δ' ἡ φύσις αὐτῶν καὶ ψηφοειδής, πλατεῖα, οὐ στρογγύλη. μέγεθος δὲ ὅσον διπλασία τῆς μεγίστης ψήφου. διαφέρει δ' αὐτῆς πρὸς τὴν δοκιμασίαν τὰ ἄνω πρὸς τὸν ἥλιον ἢ τὰ κάτω καὶ βέλτιον δοκιμάζει τὰ ἄνω· τοῦτο δὲ διότι ξηρότερα τὰ ἄνω· κωλύει γὰρ ἡ ὑγρότης εἰς τὸ ἐκλαμβάνειν· ἐπειδὴ καὶ ἐν τοῖς καύμασι δοκιμάζει χεῖρον· ἀνίησι γάρ τινα νοτίδα ἐξ αὐτῆς δι' ἣν ἀπολισθαίνει. συμβαίνει δὲ τοῦτο καὶ ἄλλοις τῶν λίθων, καὶ ἐξ ὧν τὰ ἀγάλματα ποιοῦσιν, ὃ καὶ σημεῖον ὑπολαμβάνουσιν ἴδιόν τι τοῦ εἴδους.
48 Αἱ μὲν οὖν τῶν λίθων διαφοραὶ καὶ δυνάμεις εἰσὶν ἐν τούτοις. αἱ δὲ τῆς γῆς ἐλάττονες μὲν ἰδιώτεραι δέ. τὸ μὲν γὰρ τήκεσθαι καὶ μαλάττεσθαι καὶ πάλιν ἀποσκληρύνεσθαι καὶ ταύτῃ συμβαίνει. τήκεται μὲν γὰρ †τοῖς χυτοῖς καὶ ὀρυκτοῖς ὥσπερ καὶ ὁ λίθος· μαλάττεται δέ, λίθους τε ποιοῦσιν, ὧν τάς τε ποικίλας καὶ τὰς ἄλλας συντιθεμένας † ... ἁπάσας γὰρ πυροῦντες καὶ
49 μαλάττοντες ποιοῦσιν. εἰ δὲ καὶ ὁ ὕελος ἐκ τῆς ὑελίτιδος ὥς τινές φασι, καὶ αὕτη πυκνώσει γίνεται. ἰδιωτάτη δὲ ἡ τῷ χαλκῷ μιγνυμένη· πρὸς γὰρ τῷ τήκεσθαι καὶ μίγνυσθαι καὶ δύναμιν ἔχει περιττὴν ὥστε τῷ κάλλει τῆς χρόας ποιεῖν διαφοράν. περὶ δὲ Κιλικίαν ἐστί τις ἢ ἕψεται γῆ καὶ γίνεται γλισχρά· ταύτῃ δ'
50 ἀλείφουσι τὰς ἀμπέλους ἀντὶ ἰξοῦ πρὸς τοὺς ἶπας. εἴη δ' ἂν λαμβάνειν καὶ ταύτας τὰς διαφοράς, ὅσαι πρὸς τὴν ἀπολίθωσιν εὐφυεῖς· ἐπεὶ αἵ γε τῶν τόπων ποιοῦσαι χυμοὺς διαφόρους ἀλλήλων ⟨ἰδίαν⟩ τιν' ἔχουσι φύσιν, ὥσπερ καὶ αἱ τοὺς τῶν φυτῶν. ἀλλὰ μᾶλλον ἄν τις ⟨αὐ⟩τὰς τοῖς χρώμασι διαριθμήσειεν οἶσπερ καὶ οἱ γραφεῖς χρῶνται. καὶ γὰρ ἡ γένεσις τούτων, ὥσπερ ἐξ ἀρχῆς εἴπομεν, ἤτοι συρροῆς τινὸς ἢ διηθήσεως γινομένης. καὶ ἔνιά γε δὴ φαίνεται πεπυρωμένα καὶ οἷον κατακεκαυμένα οἷον

καὶ ἡ σανδαράκη καὶ τὸ ἀρρενικὸν καὶ τὰ ἄλλα τὰ τοιαῦτα. πάντα δ' ὡς ἁπλῶς εἰπεῖν ἀπὸ τῆς ἀναθυμιάσεως ταῦτα τῆς ξηρᾶς καὶ καπνώδους. εὑρίσκεται δὴ πάντα ἐν τοῖς μετάλλοις τοῖς ἀργυ- 51 ρείοις τε καὶ χρυσείοις, ἔνια δὲ καὶ ἐν τοῖς χαλκορυχείοις, οἷον ἀρρενικόν, σανδαράκη, χρυσόκολλα, μίλτος, ὤχρα, κύανος· ἐλάχιστος δὲ οὗτος καὶ κατ' ἐλάχιστα. τῶν δ' ἄλλων τῶν μέν εἰσι ῥάβδοι, τὴν δ' ὤχραν ἀθρόαν πώς φασιν εἶναι· μίλτον δὲ παντοδαπὴν ὥστε εἰς τὰ ἀνδρείκελα χρῆσθαι τοὺς γραφεῖς· καὶ ὤχραν ἀντ' ἀρρενικοῦ διὰ τὸ μηδὲν τῇ χρόᾳ διαφέρειν, δοκεῖν δέ. ἀλλὰ μίλτου τε καὶ ὤχρας ἐστὶν ἐνιαχοῦ μέταλλα καὶ κατὰ 52 ταὐτὰ καθάπερ ἐν Καππαδοκίᾳ, καὶ ὀρύττεται πολλή. χαλεπὸν δὲ τοῖς μεταλλεῦσι φασὶν εἶναι τὸ πνίγεσθαι· ταχὺ γὰρ καὶ ἐν ὀλίγῳ τοῦτο ποιεῖν. βελτίστη δὲ δοκεῖ μίλτος ἡ Κεία εἶναι· γίνονται γὰρ πλείους. ἡ μὲν οὖν ἐκ τῶν μετάλλων, ἐπειδὴ καὶ τὰ σιδηρεῖα ἔχει μίλτον. ἀλλὰ καὶ ἡ Λημνία καὶ ἣν καλοῦσι Σινωπικήν. αὕτη δ' ἐστὶν ἡ Καππαδοκική, κατάγεται δ' εἰς Σινώπην. ἐν δὲ †τῷ μικρῷ μεταλλεύεται καθ' αὐτήν. ἔστι δὲ αὐτῆς γένη 53 τρία, ἡ μὲν ἐρυθρὰ σφόδρα, ἡ δὲ ἔκλευκος, ἡ δὲ μέση. ταύτην αὐτάρκη καλοῦμεν διὰ τὸ μὴ μίγνυσθαι, τὰς δ' ἑτέρας μιγνύουσι. γίνεται δὲ καὶ ἐκ τῆς ὤχρας κατακαιομένης ἀλλὰ χείρων, τὸ δ' εὕρημα Κυδίου. συνεῖδε γὰρ ἐκεῖνος, ὥς φασι, κατακαυθέντος τινὸς πανδοχείου τὴν ὤχραν ἰδὼν ἡμίκαυστον καὶ πεφοινιγμένην. τιθέασι δ' εἰς τὰς καμίνους χύτρας καινὰς περιπλάσαντες 54 πηλῷ· ὀπτῶσι γὰρ διάπυροι γενόμεναι· ὅσῳ δ' ἂν μᾶλλον πυρωθῶσι, τοσούτῳ μᾶλλον μελαντέραν καὶ ἀνθρακωδεστέραν ποιοῦσι. μαρτυρεῖ δ' ἡ γένεσις αὐτή· δόξειε γὰρ ἂν ὑπὸ πυρὸς ἅπαντα ταῦτα μεταβαλεῖν, εἴπερ ὁμοίαν ἢ παραπλησίαν δεῖ τὴν ἐνταῦθα τῇ φυσικῇ νομίζειν. ἔστι δέ, ὥσπερ καὶ μίλτος ἡ μὲν 55 αὐτόματος ἡ δὲ τεχνική, καὶ κύανος ὁ μὲν αὐτοφυὴς ὁ δὲ σκευαστὸς ὥσπερ ἐν Αἰγύπτῳ. γένη δὲ κυάνου τρία, ὁ Αἰγύπτιος, καὶ Σκύθης, καὶ τρίτος ὁ Κύπριος. βέλτιστος δ' ὁ Αἰγύπτιος εἰς τὰ ἄκρατα λειώματα, ὁ δὲ Σκύθης εἰς τὰ ὑδαρέστερα. σκευαστὸς δ' ὁ Αἰγύπτιος. καὶ οἱ γράφοντες τὰ περὶ τοὺς βασιλεῖς καὶ τοῦτο γράφουσι, τίς πρῶτος βασιλεὺς ἐποίησε χυτὸν κύανον μιμησάμενος τὸν αὐτοφυῆ, δῶρά τε πέμπεσθαι παρ' ἄλλων τε καὶ ἐκ Φοινίκης φόρον κυάνου, τοῦ μὲν ἀπύρου τοῦ δὲ πεπυρωμένου. φασὶ δὲ οἱ τὰ φάρμακα τρίβοντες τὸν μὲν κύανον ἐξ ἑαυτοῦ ποιεῖν χρώματα τέτταρα, τὸ μὲν πρῶτον ἐκ τῶν λεπτοτάτων λευκό-

τατον, τὸ δὲ δεύτερον ἐκ τῶν παχυτάτων μελάντατον. ταῦτά τε
56 δὴ τέχνῃ γίνεται καὶ ἔτι τὸ ψιμύθιον. τίθεται γὰρ μόλυβδος
ὑπὲρ ὄξους ἐν πίθοις ἡλίκον πλίνθος. ὅταν δὲ λάβῃ πάχος, λαμ-
βάνει δὲ μάλιστα ἐν ἡμέραις δέκα, τότ' ἀνοίγουσιν, εἶτ' ἀπο-
ξύουσιν ὥσπερ εὐρῶτά τινα ἀπ' αὐτοῦ, καὶ πάλιν, ἕως ἂν κατανα-
λώσωσι. τὸ δ' ἀποξυόμενον ἐν τριπτῆρι τρίβουσι καὶ ἀφηθοῦσιν
57 ἀεί, τὸ δ' ἔσχατον ὑφιστάμενόν ἐστι τὸ ψιμύθιον. παραπλησίως
δὲ καὶ ὁ ἰὸς γίνεται· χαλκὸς γὰρ ἐρυθρὸς ὑπὲρ τρυγὸς τίθεται
καὶ ἀποξύεται τὸ ἐπιγινόμενον αὐτῷ· ἐπιφαίνεται γὰρ ὁ ἰός.
58 γίνεται δὲ καὶ κιννάβαρι τὸ μὲν αὐτοφυὲς τὸ δὲ κατ' ἐργασίαν.
αὐτοφυὲς μὲν τὸ περὶ Ἰβηρίαν σκληρὸν σφόδρα καὶ λιθῶδες,
καὶ τὸ ἐν Κόλχοις. τοῦτο δέ φασιν εἶναι ⟨ἐπὶ⟩ κρημνῶν ὃ κατα-
βάλλουσι τοξεύοντες. τὸ δὲ κατ' ἐργασίαν ὑπὲρ Ἐφέσου μικρὸν
ἐξ ἑνὸς τόπου μόνον. ἔστι δ' ἄμμος ἣν συλλέγουσι λαμπυρί-
ζουσαν καθάπερ ὁ κόκκος· ταύτην δὲ τρίψαντες ὅλως ἐν ἀγγείοις
λιθίνοις λειοτάτην πλύνουσιν ἐν χαλκοῖς [μικρὸν ἐν καλοῖς]
τὸ δ' ὑφιστάμενον πάλιν λαβόντες πλύνουσι καὶ τρίβουσιν, ἐν
ᾧπερ ἐστὶ τὸ τῆς τέχνης· οἱ μὲν γὰρ ἐκ τοῦ ἴσου πολὺ περι-
ποιοῦσιν, οἱ δ' ὀλίγον ἢ οὐθέν· ἀλλὰ πλύσματι ⟨τῷ⟩ ἐπάνω
χρῶνται ἓν πρὸς ἓν ἀλείφοντες. γίνεται δὲ τὸ μὲν ὑφιστάμενον
59 κάτω κιννάβαρι, τὸ δ' ἐπάνω καὶ πλεῖον πλύσμα. καταδεῖξαι δέ
φασι καὶ εὑρεῖν τὴν ἐργασίαν Καλλίαν τινὰ Ἀθηναῖον ἐκ τῶν
ἀργυρείων, ὃς οἰόμενος ἔχειν τὴν ἄμμον χρυσίον διὰ τὸ λαμ-
πυρίζειν ἐπραγματεύετο καὶ συνέλεγεν. ἐπεὶ δ' ᾔσθετο ὅτι οὐκ
ἔχει τὸ δὲ τῆς ἄμμου κάλλος ἐθαύμαζε διὰ τὴν χρόαν οὕτως ἐπὶ
τὴν ἐργασίαν ἦλθε ταύτην. οὐ παλαιὸν δ' ἐστὶν ἀλλὰ περὶ ἔτη
60 μάλιστ' ἐνενήκοντα εἰς ἄρχοντα Πραξίβουλον Ἀθήνησι. φανερὸν δ' ἐκ τούτων ὅτι μιμεῖται τὴν φύσιν ἡ τέχνη, τὰ δ' ἴδια ποιεῖ,
καὶ τούτων τὰ μὲν χρήσεως χάριν τὰ δὲ μόνον φαντασίας ὥσπερ
τὰς †ἄλπεις. ἔνια δὲ ἴσως ἀμφοῖν ὥσπερ χυτὸν ἄργυρον. ἔστι
γάρ τις χρεία καὶ τούτου. ποιεῖται δὲ ὅταν τὸ ⟨κιννάβαρι⟩ τριφθῇ
μετ' ὄξους ἐν ἀγγείῳ χαλκῷ καὶ δοίδυκι χαλκῷ. τὰ μὲν οὖν
τοιαῦτα τάχ' ἄν τις λάβοι πλείω.
61   Τῶν δὲ μεταλλευτῶν τὰ ἐν τοῖς γεωφανέσιν ἔτι λοιπά, [περὶ]
ὧν ἡ γένεσις ὥσπερ ἐλέχθη κατ' ἀρχὰς ἐκ συρροῆς τινος καὶ
ἐκκρίσεως γίνεται καθαρωτέρας καὶ ὁμαλωτέρας τῶν ἄλλων.
χρώματα δὲ παντοῖα λαμβάνουσι καὶ διὰ τὴν τῶν ὑποκειμένων
† ... διὰ τὴν τῶν ... ουντων διαφορὰν, ἐξ ὧν τὰς μὲν μαλάτ-

τοντες, τὰς δὲ τήκοντες καὶ τρίβοντες συντιθέασι τὰς λίθους τὰς ἐκ τῆς Ἀσίας ταύτας ἀγομένας. αἱ δ' αὐτοφυεῖς καὶ ἅμα 62 τῷ περιττῷ τὸ χρήσιμον ἔχουσαι σχεδὸν τρεῖς εἰσὶν ἢ τέτταρες, ἥ τε Μηλιὰς καὶ ἡ Κιμωλία καὶ ἡ Σαμία καὶ ἡ Τυμφαϊκὴ τετάρτη παρὰ ταύτας ἢ γύψος. χρῶνται δὲ οἱ γραφεῖς τῇ Μηλιάδι μόνον, τῇ Σαμίᾳ δ' οὔ, καίπερ οὔσῃ καλῇ, διὰ τὸ λίπος ἔχειν καὶ πυκνότητα καὶ λειότητα. τὸ γὰρ †ἤρεμον καὶ ... δες καὶ ἀλιπὲς ἐπὶ τῆς γραφῆς ἁρμόττει μᾶλλον ὅπερ ἡ Μηλιὰς ἔχει † τῷ φαρίδι. εἰσὶ δὲ ἐν τῇ Μήλῳ καὶ ἐν τῇ Σάμῳ διαφοραὶ τῆς γῆς 63 πλείους. ὀρύττοντα μὲν οὖν οὐκ ἔστιν ὀρθὸν στῆναι ἐν τοῖς ἐν Σάμῳ ἀλλ' ἀναγκαῖον ἢ ὕπτιον ἢ πλάγιον. ἡ δὲ φλὲψ ἐπὶ πολὺ διατείνει, τὸ μὲν ὕψος ἡλίκη δίπους, τὸ δὲ βάθος πολλῷ μείζων· ἐφ' ἑκάτερα δ' αὐτὴν λίθοι περιέχουσιν ἐξ ὧν ἐξαιρεῖται. διαφυὴν ἔχει διὰ μέσου καὶ ἡ διαφυὴ βελτίων ἐστὶ τῶν ἔξω καὶ πάλιν ἑτέραν αὐτῆς καὶ ἑτέραν ἄχρι τεττάρων † ... ἐστὶν ἡ ἐσχάτη, καλεῖται ἀστήρ· χρῶνται δὲ τῇ γῇ πρὸς τὰ ἱμάτια μά- 64 λιστα ἢ μόνον. χρῶνται δὲ καὶ τῇ Τυμφαϊκῇ πρὸς τὰ ἱμάτια καὶ καλοῦσι γύψον οἱ περὶ τὸν Ἄθων καὶ τοὺς τόπους ἐκείνους. ἡ δὲ γύψος γίνεται πλείστη μὲν ἐν Κύπρῳ καὶ περιφανεστάτη. μικρὸν γὰρ ἀφαιροῦσι τῆς γῆς ὀρύττοντες. ἐν Φοινίκῃ δὲ καὶ ἐν Συρίᾳ καίοντες τοὺς λίθους ποιοῦσιν. ἔπειτα δ' ἐν Θουρίοις· καὶ γὰρ ἐκεῖ γίνεται πολλή. τρίτη δὲ ἡ περὶ Τυμφαίαν καὶ περὶ Περραιβίαν καὶ κατ' ἄλλους τόπους. ἡ δὲ φύσις αὐτῆς ἰδία· 65 λιθωδεστέρα γὰρ μᾶλλόν ἐστιν ἢ γεώδης· ὁ δὲ λίθος ἐμφερὴς τῷ ἀλαβαστρίτῃ· μέγας δ' οὐ τέμνεται ἀλλὰ χαλικώδης. ἡ δὲ γλισχρότης καὶ θερμότης ὅταν βρεχθῇ θαυμαστή. χρῶνται γὰρ πρός τε τὰ οἰκοδομήματα τὸν λίθον περιχέοντες κἄν τι ἄλλο βούλωνται τοιοῦτο κολλῆσαι. κόψαντες δὲ καὶ ὕδωρ ἐπιχέοντες 66 ταράττουσι ξύλοις, τῇ χειρὶ γὰρ οὐ δύνανται διὰ τὴν θερμότητα. βρέχουσι δὲ παραχρῆμα πρὸς τὴν χρείαν· ἐὰν ⟨δὲ⟩ μικρὸν πρότερον ταχὺ πήγνυται καὶ οὐκ ἔστι διελεῖν. θαυμαστὴ δὲ καὶ ⟨ἡ⟩ ἰσχύς· ὅτε γὰρ οἱ λίθοι ῥήγνυνται ἢ διαφέρονται ἡ γύψος οὐκ ἀνίησι, πολλάκις δὲ καὶ τὰ μὲν πέπτωκε καὶ ὑφῄρηται, τὰ δ' ἄνω κρεμάμενα μένει συνεχόμενα τῇ κολλήσει. δύναται δὲ 67 καὶ ὑφαιρουμένη πάλιν καὶ πάλιν ὀπτᾶσθαι καὶ γίνεσθαι χρησίμη. περὶ μὲν οὖν Κύπρον καὶ Φοινίκην εἰς ταῦτα μάλιστα, περὶ δὲ Ἰταλίαν καὶ εἰς τὸν οἶνον· καὶ οἱ γραφεῖς ⟨εἰς⟩ ἔνια τῶν κατὰ τὴν τέχνην ἔτι δὲ οἱ γναφεῖς ἐμπάττοντες εἰς τὰ ἱμάτια.

διαφέρειν δὲ δοκεῖ καὶ πρὸς τὰ ἀπομάγματα πολὺ τῶν ἄλλων, εἰς ὃ καὶ χρῶνται μᾶλλον καὶ μάλισθ' οἱ περὶ τὴν Ἑλλάδα, γλισ-
68 χρότητι καὶ λειότητι. ἡ μὲν δύναμις ἐν τούτοις καὶ τοῖς τοιούτοις. ἡ δὲ φύσις ἔοικεν ἀμφότερά πως ἔχειν καὶ τὰ τῆς κονίας καὶ τὰ τῆς γῆς, θερμότητα καὶ γλισχρότητα, μᾶλλον δὲ ἑκατέραν ὑπερέχουσαν. ὅτι δ' ἔμπυρος κἀκεῖθεν φανερόν. ἤδη γάρ τις ναῦς ἱματηγὸς βρεχθέντων ἱματίων ὡς ἐπυρώθησαν συγκατεκαύθη
69 καὶ αὐτή. καίουσι δὲ καὶ ἐν Φοινίκῃ καὶ ἐν Συρίᾳ καμινεύοντες αὐτήν [καὶ καίοντες]· καίουσι δὲ μάλιστα τοὺς μαρμάρους καὶ †ἁπλουστέρους, στερεωτάτους μὲν παρατιθέντες ⟨βόλιτον, ἕνεκα⟩ τοῦ θᾶττον καίεσθαι καὶ μᾶλλον. δοκεῖ γὰρ θερμότατον εἶναι πυρωθὲν καὶ πλεῖστον χρόνον διαμένει. ὀπτήσαντες δὲ κόπτουσιν ὥσπερ τὴν κονίαν. ἐκ τούτου δ' ἂν δόξειεν εἶναι φανερὸν ὅτι πυρώδης τις ἡ γένεσις αὐτῆς τὸ ὅλον ἐστίν.

APPARATUS CRITICUS

1 μεταλλευόμενα/ μεταλευόμενα B,C.
εἴδη/ om. Furlanus; restored by Schneider.
χρώμασιν/ Aldus; χρώμενα Ω.
ἄλλῃ/ Schneider; καὶ ἄλλῃ Ω,Aldus.
μεταλλευομένων/ μεταλευομένων Ω.
τούτων/ τῶν λίθων conj.Wimmer.
λέγωμεν/ Aldus; λέγομεν Ω.
2 οὖν/ Aldus; ἂν Ω.
συρροῆς/ conj.Schneider,Wimmer in text (cf. §50); ῥοῆς Ω,Aldus.
διηθήσεώς τινος/ Wimmer; δ. διά τινος Ω,Aldus; δ. [διά] τινος Schneider.
γενομένης/ Wimmer; γινομένης A,C,Aldus.
ὡς ἀνωτέρω/ Aldus; ἀνωτέρως Ω.
καὶ κατ'/ καὶ om.C.
ὁμαλέστερον/ Schneider; καὶ ὁμαλεστέρων Ω; καὶ ὁμαλέστερον Aldus.
καθαρώτερον/ καθαρωτέρων Ω.
τὰ κατὰ/ Schneider; τε κατὰ Ω,Aldus.
3 ἐπεὶ τά γε/ Turnebus, Schneider; ἐ. τά τε Ω,Aldus.
δόξειεν ἂν/ Turnebus, Schneider; ἂν om.Ω,Aldus.
ἐπείπερ/ Schneider; ἐ. ἂν A,C,Aldus; ἐ. δὴ B; ἐ. ἐν Turnebus.
ἐν μὲν/ Turnebus, Schneider; μὲν om.Ω,Aldus.
πολλαὶ διαφοραί/ Turnebus, Schneider; πνοαὶ διάφοροι Ω,Aldus; ῥοαὶ διάφοροι Furlanus.
4 αὐταί τε/ Turnebus; ἅπτεται Ω,Aldus.
ταύταις/ Turnebus; ταῖς ἑαυταῖς Ω,Aldus.
τοῦ τε ποιεῖν/ Turnebus; οὔτε π. Ω,Aldus.
τηκτοὶ γὰρ/ οἱ μὲν γὰρ τ. Turnebus, Schneider.
καυστοὶ/ οἱ μὲν κ. Turnebus.
ἐξομοιοῦν[ται]/ ἐξομοιοῦνται Ω,Aldus; ἐξομοιοῦν φαίνονται Turnebus; ἐξομοιοῦν λέγονται Furlanus,Wimmer.
τὸν χρυσὸν καὶ/ om.Heinsius; restored by Schneider.
5 τικτόντων/ Schneider; τικτῶν Ω,Aldus; τηκτῶν Turnebus.
[τῶν] καὶ/ Schneider; τῶν κ. A,C,Aldus; καὶ τῶν B.
⟨ἡ⟩ κατὰ τὰς ἐργασίας/ conj.Schneider,Stephanides; ἡ om.Ω,Wimmer. Possibly ⟨ἡ τῶν⟩.
ἄλλαι/ Turnebus; ἄλλα Ω,Aldus.
παρὰ ταύτας/ Turnebus, Schneider; κατὰ τ. Ω,Aldus; κατὰ ταύτας ἰδιότητας Hill.
διαφοραί/ om.Ω,Aldus; ⟨διαφοραί⟩ Furlanus; κατὰ ταύτας ἰδιότητας διαφοραί conj.Laet.
6 δι' ὧν τὸ περιττόν/ Ω; διὸ τὸ π. Aldus,Turnebus; διὰ τὸ π. Furlanus, Wimmer.

· 31 ·

ὅλον/ ὅλου Ω,Aldus.
Θηβαϊκῶν/ Aldus; θηραικῶν Ω.
ὁ περὶ/ ὁ om.B.
περὶ Αἴγυπτον ἐν Θήβαις/ ἐν Αἰγύπτῳ περὶ Θήβας Hill.
ἀλαβαστρίτης/ Aldus; λαβαστρίτης A,C; B(corr. ἀλ-).
ὅμοιος/ μόνος A,C,Aldus; B(corr. ὅμοιος).
πυέλῳ/ πέπλω Ω,Aldus, πέπλῳ Turnebus; πυέλῳ Laet (conj.Salmasius, P.E.848aC).

7 πόρος/ πῶρος Laet,Hill.
πυκνότητι/ πυκνώτητι A.
καὶ μέλας/ Wimmer; καὶ om.Ω,Aldus; εὑρίσκεται καὶ μ. Hill; ⟨ἔστι δέ τις καὶ⟩ μ. Schneider.
ὅμοιος/ Schneider; ὁμοίως A,C,Aldus; B ὁμοίῳ, corr. -ος.
πλείους/ B,Furlanus; πλει... A,C,Aldus.
μὲν οὖν/ οὖν om.B.
τόποις ὅλοις/ conj.Schneider,Wimmer in text; τοῖς ὅλοις Ω,Aldus.

8 ἔνιοι/ Aldus; ἐνίοις Ω.
σπάνιοι/ σπανίοις C.
σχεδὸν ⟨πάντες τῶν κατὰ⟩ λόγον εἰς τὰ σφραγίδια γλυπτῶν/ πάντες τῶν κατὰ om.Ω,Wimmer; σχεδὸν λόγῳ τῶν ἐ.τ.σ.γ. Laet (conj.Salmasius, P.E.69aD) and Hill; σχεδὸν ὅσοι καταλέγονται ἐ.τ.σ. τῶν γ. conj.Schneider; σχεδὸν πάντες οἱ κατὰ λόγον ἐ.τ.σ. γλυπτοί conj.Stephanides.

9 σιδήρῳ/ ⟨τῷ⟩ σιδήρῳ Schneider.
εἴτ' οὖν/ Schneider; εἰ τοίνυν Ω,Aldus.
ἐνυπαρχόντων/ Schneider; ὑπαρχόντων Ω,Aldus.
δι' αὐτούς/ Schneider; δι' αὐτάς Ω (αὐτάς A), Aldus.
μυλίαι/ Turnebus; μιλίαι Ω,Aldus.
ῥέουσιν οἷς/ ῥ. σὺν οἷς conj.Schneider; possibly συρρέουσιν οἷς.
οἱ δὲ/ εἰ δὲ B.
πάντας/ πάντα Ω.

10 οὕτως/ Wimmer; οὕτως ὅλως Ω,Aldus.
εἰρῆσθαι/ ῥῆσθαι B.
διαπηδῶντες/ διαπηδόντες B.
ἀπομαχόμενοι/ Schneider; οὐ μαχομένου Ω,Aldus; οὐ μαχόμενοι Furlanus.
τὴν πύρωσιν/ ⟨κατὰ⟩ τ.π. Furlanus.
[οὐδ']/ Schneider; οὐδ' Ω,Aldus.
ἐξυγρασμένοι/ Aldus; ἐξηγορασμένοι Ω.
τὸ γὰρ/ Aldus; ὁ γὰρ Ω.
τηκτὸν/ τικτὸν C.
δεῖ/ Schneider; ἀεὶ Ω,Aldus.
ἔχειν/ Schneider; ἔχει Ω,Aldus.

11 καταβρεχθέντας/ Furlanus; καταρρηχθέντας A,C; κατ' ἀρηχθέντας B.
συνικμασθέντας/ συνεκμασθέντας B,Aldus.
ὧν/ Turnebus; ὡς Ω,Aldus.

APPARATUS CRITICUS

τηκτούς/ τικτούς C.
12 καύσει/ Hill; θραύσει Ω,Aldus.
οὖς/ καὶ οὖς Hill.
ἄχρις ἂν/ Turnebus, Schneider; χρειαν Ω,Aldus (various accents); χρείας Furlanus; χρείας ἐὰν Hill.
ὀσμὴ/ Turnebus; ὡς μὴ Ω,Aldus.
13 ὃν δὲ/ ὃν Aldus; ὃν ⟨δὲ⟩ Furlanus.
σπίνον/ σπῖνον Wimmer.
ἐν τοῖς/ τοῖς ⟨αὐτοῖς⟩ Wimmer.
οὗτος/ Schneider; τοιοῦτος Ω,Aldus.
14 ἐκφοροῦται τε/ τε om.B; ἐκπωροῦται Laet (also in sec.15).
κισσηροειδὴς/ B,Turnebus; κισηροειδὴς A, C, Aldus.
ὥσθ'/ ὡς ἐθ' B.
χρόαν/ χρείαν B.
κυττάρῳ/ Schneider; κυθρισμῶ A,B, κυθρρισμῶ C, κυθρισμῷ Aldus; κυτταρείῳ Furlanus.
ἐκεῖνος μὲν/ Furlanus; εἰ μαν Ω, εἰ μὲν Aldus; ἐν μὲν Turnebus.
λιπαραίῳ/ Aldus; λιπαρῶ Ω.
15 Τετράδι/ Aldus; τεταρίδι Ω.
Λιπάραν/ λιπάρας B,C.
καλουμένῃ/ καλουμένης B.
τῷ ἐν Βίναις/ Wimmer; ταῖς κίναις Ω,Aldus; τοῖς ἐν Βίναις Turnebus; ταῖς Βίναις Furlanus; ⟨τῷ ἐν⟩ ταῖς Βίναις Schneider.
κατακαύσεως/ Turnebus; κατακλίσεως Ω,Aldus.
γῇ/ Turnebus; τῇ Ω, τῇ Aldus.
16 ὀρυττομένων/ Schneider(Syll.); θρυττομένων Ω, θριτομένων Aldus, θρυπτομένων Turnebus.
χρείαν/ Aldus; χρόαν Ω.
ἐκκαίονται/ ἐκκαίοντε C.
Λιγυστικὴν/ Furlanus; λυγιστικὴν Ω,Aldus.
Ἠλείᾳ/ Furlanus; ἰλία Ω, ἰλίᾳ Aldus; ἤλιδι Turnebus.
Ὀλυμπίαζε/ ὀλιμπίαζε B.
17 ἐν τοῖς Σκαπτησύλης/ ἐγκαπτῆς ὕλης Ω,Aldus; ἐν σκαπτησύλης Turnebus; ἐν ⟨τοῖς⟩ σ. Furlanus.
παρόμοιος/ παρόμοι B.
ἐκκαίεται/ Ω; καίεται Aldus,Wimmer; perhaps ἐκαίετο.
τότε παύεται/ Ω; perhaps τότ' ἐπαύετο.
ἀπαθὴς ὤν/ Furlanus; ἀ. ἦν Ω,Aldus.
18 ὅλως/ ὡς ὁ Turnebus.
ἐρυθρὸν/ ἐρυθρῶ B.
Μασσαλίας/ Μασαλίας B.
σφόδρα/ σφοδρὸν Aldus; σ. ὢν Turnebus.
19 καίεται δὲ ὁ/ Aldus; κ. δύο Ω.
γωνιοειδὴς/ γωνιωειδὴς B, γωνιειδὴς Aldus, γωνιώδης Turnebus.

· 33 ·

κίσσηρις (before καὶ)/ Schneider; κίτηρις Ω, κίττηρις Aldus; but κίσσηρις before ἐκ.
δόξειεν/ Turnebus; δόξειε δ' Ω,Aldus.
20 γινομένων/ Schneider; γενομένων Ω,Aldus.
διαβάρου/ 'Αραβικοῦ Hill ('Αρ.); perhaps διαβόρου.
ἡ κισσηροῦται/ B,C,Turnebus; ἡ κ. A; ἡ καὶ κ. Furlanus, Heinsius, Laet, Hill,Schneider; οὐ κ. Aldus,Wimmer.
ἐν τοῖς/ ἐν ταῖς Ω,Aldus; ἐν Αἴτνῃ conj.Salmasius (cf.Schneider IV, p. 554), ἐν τοῖς ⟨καιομένοις⟩ Schneider; perhaps ἐν τοῖς πυρικαύστοις, or ἐν τούτοις, or ἐν τοῖς τοιούτοις.
21 κατὰ μικρὰ/ Perhaps κατὰ μικρὰς or καὶ μικρὰς, Coraës,Strabo IV, p. 116, on Strabo VII,299 (cf.Schneider,Syll.).
χειροπληθεῖς/ Turnebus; χειροπλιστίας A,B; χειροπλησίας C,Aldus.
ἀπαμήσωνται/ Schneider; ἀπαμείβωνται Ω,Aldus; ἀπαμῶνται Turnebus; ἀπαμείρωνται Furlanus,Coraës.
ἡ ἄμμος/ Schneider; καὶ ἡ ἀ. Ω,Aldus.
αὖ καὶ/ Aldus; αὖ Ω.
ἡ ἄμμος. ἡ δ' αὖ καὶ/ καὶ ἀμμώδης Hill (conj.Laet).
πᾶσα μὲν ... ἔνια δ' αὖ ἐν λίθῳ/ π. μ. εὔθραυστος, ἐν λίθῳ δὲ conj.Schneider; π. μ. σχεδὸν ὡς ἐν Νισύρῳ, ἔνια δ' αὖ ἐ. λ. conj.Stephanides.
ἑτέρῳ/ ἑτέρᾳ Aldus.
22 μέλαινα ⟨ἡ⟩ ἐκ τοῦ ῥύακος/ conj.Stephanides; ἡ om.Ω,Wimmer.
πυκνότητι δὲ καὶ βάρει/ π. τε κ. β. Ω,Aldus; πυκνός τε καὶ βαρεῖα Furlanus.
αὕτη/ Schneider; αὐτή Ω,Aldus.
⟨ἡ⟩ μαλώδης/ ἡ om.Ω,Wimmer; ἁλμώδης Turnebus; μυλώδης Furlanus; ἡ Μηλία conj.Schneider; ἡ μυλώδης conj.Stephanides. For ἡ μηλώδης compare ὑποκίτρινος, Stephanides (Min.).
πολυτιμοτέρα/ A,C; πολυτιμότερος B; πολυτιμότερον Aldus,Wimmer.
ἑτέρας/ ἑταίρας C.
τμητικὴ ... τμητικωτάτη/ σμηκτικὴ ... σμηκτικωτάτη Hill.
ἡ ἐκ τοῦ/ ἡ om.B.
⟨ἡ⟩ ἐκ τῆς θαλάσσης/ conj.Stephanides; ἡ om.Ω,Wimmer.
ῥύακος (before μᾶλλον)/ ῥικὸς A,B,Aldus; ? ῥινὸς C. But ῥύακος above before τοῦ.
κούφης/ Furlanus; κορυφῆς Ω,Aldus.
αὐτῆς/ αὐτοῦ Aldus.
τῆς κισσήριδος/ Ω,Aldus; τῆς om.Wimmer.
23 ⟨διάφοροι⟩/Wimmer; om.Ω,Aldus; διαφοραὶ Turnebus; κατὰ τὰς ἰδιότητας διαφοραὶ Hill (conj.Laet).
αἱ μὲν τῇ ὄψει μόνον/ αἱ μὲν τῇ ὄψει ⟨διαφέρουσαι⟩, ὁμώνυμοι ⟨δὲ πρὸς ἀλλήλας⟩ conj.Schneider (cf.§30).
ἡ ἴασπις/ ἡ ἄσπις Ω; ἴασπις Aldus.
μὲν οὖσα/ μὲν οὖν οὖσα B.
παντὸς/ Turnebus; πάντως Ω,Aldus.

APPARATUS CRITICUS

24 φορούσιν/ Aldus; διαφορούσιν Ω.
βλέπειν/ Perhaps εὖ βλέπειν.
ἐκείνοις/ conj.Schneider,Wimmer in text; . . . νους A,C,Aldus; σμαράγδους B (σμαρά- in later hand); ἔνιοι Furlanus.
ἐκείνοις γάρ φασι/ φασὶ γὰρ Hill.
κομισθῆναι/ Aldus; κοσμηθῆναι Ω.
ὀβελίσκῳ/ Aldus; ὀβελίσκους Ω,Schneider, corr. ὀβελίσκῳ Syll.; ὀβέλισκον Hill.
σμαράγδους/ σμαράγδου Schneider.
σμαράγδους τέτταρας/ ἐκ σμαράγδων τεττάρων Hill.
μὲν οὖν/ οὖν om.B.
25 τανῶν/ Turnebus,Hill,Stephanides; . . . ανῶν Ω,Aldus; βακτριανῶν Furlanus,Heinsius,Schneider,Wimmer.
καλουμένων/ Turnebus; καιομένων A,C,Aldus,B (corr. λεω above line).
Ἡρακλέους/ ἱρακλέους A.
τις φύσις/ τι φ. B.
τόποις/ τοποι B.
ἐν τοῖς χαλκορυχείοις/ ἐν τοῖς αὐτοῖς χ. A.
χαλκορυχείοις/ χαλκωρυχείοις Turnebus; also in §§26,51.
τῇ νήσῳ/ Δημονήσῳ conj.Salmasius(P.E.505aD).
Χαλκηδόνι/ Schneider (conj.Salmasius,137bD); καρχηδόνι Ω,Aldus.
ἰδιωτέρους/ ἰδιωτέροις B.
ἐποίησεν/ Wimmer; ποιοῦσιν Ω,Aldus.
26 παρόμοιαι/ Schneider; παρόμοια Ω,Aldus.
χρυσείοις/ Aldus; χρυσίοις Ω.
περὶ τοὺς/ om.B.
περὶ τοὺς . . . τόπους/ π. τ. στόβους Turnebus (Στόβους Schneider); περὶ τοὺς Καταδούπων τόπους conj.Stephanides (cf.§34).
27 Κύπρῳ/ Aldus; Κύθρω Ω.
ἀργὴ/ Schneider; ἀρχὴ Ω,Aldus.
28 αὔτη/ Laet; αὐτή Ω,Aldus.
λυγγούριον/ λυγκούριον Laet; see also §31.
ξύλον/ φύλλα conj.Wimmer.
ἔστι δὲ/ Schneider; ἔτι δὲ Ω; ἔτι Aldus.
διαφανές/ Schneider; διαφανῆ Ω,Aldus (-ή).
σφόδρα καὶ/ καὶ σ., corr. σ. κ. C.
ψυχρόν/ Schneider; ψυχρά Ω,Aldus; πυρρά Furlanus.
τὸ τῶν (four times)/ Schneider; τὰ τῶν Ω,Aldus.
καὶ τῆς τοῦ σώματος ὅλως φύσεως, ᾗ τὸ μὲν ξηρότερον/ From Ω (but ᾖ Ω, τὸ μὲν om.Ω); om.Aldus,Turnebus,Schneider;add.Schneider (Syll.); ⟨καὶ . . . ξηρότερον⟩ Furlanus, Heinsius, Laet.
ἔμπειροι/ ἔμπυροι A.
οὐρήσῃ/ Turnebus; εὑρήσει A,B; εὑρήσῃ C,Aldus.
29 ἐπεὶ δὲ/ ἔπειτα Schneider.
τὸ γὰρ/ καὶ γὰρ Furlanus.

· 35 ·

ὃ περὶ Λιγυστικήν/ ὃ ⟨γίνεται⟩ περὶ ⟨τὴν⟩ Λ. Schneider.
ὃ περὶ/ τὸ περὶ Furlanus.
Λιγυστικήν/ Laet; λυγγιστήν Ω,Aldus.
τούτῳ/ Aldus; τούτων Ω.
ἀκολουθοίη/ ἀκολουθείη B,C,Aldus.
ἐπίδηλος/ Wimmer; ὅτι δῆλον Ω,Aldus; ὅτι δῆλος Heinsius.
ἡ τὸν σίδηρον ἄγουσα/ Wimmer; τὸν σ. ἄγουσι Ω,Aldus; τὸν σ. ἄγουσα Heinsius.
30 κρύσταλλος/ κρύσταλος C.
καλεῖται/ Aldus; καὶ κ. Ω.
[καὶ] ἄρσεν/ Schneider; καὶ ἀ. Ω,Aldus.
31 καλεῖται δὲ/ καὶ κ. δ. B.
ἄρρην/ Laet; ἄρρεν Ω,Aldus.
θῆλυς/ Laet; θῆλυ Ω,Aldus.
ὁ ἄρρην/ A; ὁ ἄρρεν B,C,Aldus.
μικτὸν/ Wimmer; μικτὴ Ω,Aldus.
οἰνωπὸν/ οἰνοπὸν B.
Ἀχάτου/ ὀχάτου A.
32 ἀνενεχθείσης/ Aldus; ἀναινεχθείσης Ω.
στιρὰν/ A,B,Aldus; στιρρὰν C; Τίραν Turnebus,Laet,Hill; Τύραν conj. Laet; σφύραν Furlanus,Heinsius; Ἄστυρα conj.Schneider; (Ἄστυρα) add.Wimmer; Στάτειραν Highet.
γλυφθὲν/ Turnebus,Schneider; γλυφερὸν Ω,Aldus.
ἀνεπέμφθη/ Ἀλεξάνδρῳ ἐπέμφθη Turnebus; ἀνεπέμφθη (Ἀλεξάνδρῳ) Schneider,Wimmer.
33 ἐξ Ὀρχομενοῦ/ Turnebus; ἐξερχόμενον A,C,Aldus; B (corr. ὀρχομένου above line).
κάτοπτρα δὲ/ καθοὶ A,C; καθὸ B,Aldus; κάτοπτρα Turnebus.
ὅτι χλωροειδέστερος/ Schneider; τὸ ... χλωροειδέστερον A,C,Aldus; B(corr. λευκὸν after τὸ in another hand); τὸ λευκότερον καὶ χ. Hill.
34 Μασσαλίαν/ Heinsius; μασαλίαν Ω,Aldus.
κατὰ τοὺς Καταδούπους/ Schneider; καὶ τοὺς κατάδου τόπους Ω,Aldus; καὶ τῶν καταδούπων Turnebus; καὶ ἐκ τῶν κ. Furlanus.
Συήνης/ Turnebus; συήνη Ω,Aldus.
ψεφὼ/ ψεβὼ conj.Salmasius (P.E.269aG); ψηβὼ Hill.
35 ἴασπις/ ἄσπις B.
λιθοκόλλητα/ Schneider; λιθόκολλα Ω,Aldus.
τῆς Βακτριανῆς/ τοῖς βακτριανοῖς B.
τοὺς ἐτησίας/ A,B; τοὺς ἐτησίους (corr. ? -ας) C; τοὺς om.Aldus, Turnebus,Wimmer.
κινουμένης/ κενουμένης A.
36 διαφανὴς/ οὐ διαφανὴς Hill (conj.Salmasius,P.E.784aA).
τοὺς πολυτελεῖς/ Ω,Aldus,Schneider,Wimmer(T); τοὺς om.Wimmer (D).

APPARATUS CRITICUS

⟨πλὴν ἐλάττονι· μέγεθος δὲ ἡλίκος ἰχθύος ὀφθαλμὸς εὐμεγέθης⟩/ om.Ω, Aldus; add.Schneider from Athenaeus (3,93) with ἡλίκος for ἡλίκον.
ταύταις/ Schneider; αὐταῖς Ω,Aldus.
37 ὁ ἐλέφας/ Schneider; ὅ τε ἀλάφας A,C; B (corr. from ἀλάφορας); ὅ τε ἐλέφας Aldus.
ὀρυκτός/ Turnebus; ὀρεκτὸς Ω,Aldus.
καὶ λευκῷ/ om.Aldus,Turnebus; ⟨καὶ λευκῷ⟩ Furlanus; restored by Schneider.
πρασῖτις ... αἱματῖτις/ πρασίτις ... αἱματίτις Wimmer; πρασίτης Hill.
καὶ κατὰ/ Schneider; ἡ κατὰ Ω,Aldus.
ἄλλη δὲ ἡ/ ἅ. δ' ἡ A; ἅ. δὴ C,Aldus; ἅ. δὲ B.
καλουμένη ξανθή/ Hill,Schneider; ξανθή om.Ω,Aldus.
οὐ ξανθὴ/ ξανθὴ om.Hill.
μᾶλλον ὃ/ Schneider; ὃ μᾶλλον Ω,Aldus.
Δωριεῖς/ Schneider; δωρεῖς Ω,Aldus.
38 κουράλιον/ κουράλλιον Hill.
ὡς ῥίζα/ Wimmer; ὡς ἂν ῥ. A,B,Aldus; C(ὡσὰν).
39 μεταλλευομένων/ μεταλευομένων C.
ἔνιαι/ ἔνια Ω,Aldus; ἔνιοι Turnebus.
βάρος δὲ ἔχουσι/ β. δ. ἔχουσα Schneider.
40 τὸ δὲ ὅλον/ Schneider; δὲ om.Ω,Aldus; τὸ ὅλον δὲ Heinsius.
τὰ δὲ οἷον/ Furlanus; τὸ δὲ οἷον A,C,Aldus; τὸ δ' ὧν B.
σανδαράκη/ σανδράκη C.
41 σιδηρίοις/ σιδήροις B.
πριστοὶ/ ἄλλοι πριστοὶ Heinsius.
πριστοὶ γάρ/ οἱ μὲν γὰρ π. Schneider.
μαγνῆτις/ μαγνήτης B.
ὄψει περιττὸν/ τῇ ὄψει τὸ π. Schneider.
ἧς γε ... οὔσης συγγενοῦς/ conj.Schneider,Wimmer in text; ὥς γε ... οὖσαν συγγενῆ Ω,Aldus.
42 ὀρυκτὸς/ Turnebus; ὀρεκτὸς Ω,Aldus.
ὡς/ ὃς Heinsius.
βωλώδης/ Schneider; βόλασος Ω,Aldus; βολώδης Furlanus.
ὅταν δὲ/ δὲ om.Aldus; ⟨δὲ⟩ Furlanus.
καὶ ἀποβαφῇ/ om.Aldus; ⟨καὶ ἀ.⟩ Furlanus.
τῷ ἐλαίῳ/ Ω,Aldus; [τῷ] ἐ. Wimmer.
43 οἱ μὲν ⟨οὖν⟩/ Schneider; οὖν om.Ω,Aldus.
προσδέχονται/ ὑποδέχονται Heinsius.
σιδήροις δ' οὐ δύνανται/ σίδηρος δ' οὐ δύναται Hill.
οἱ δὲ/ οὐδὲ B.
ἀμβλυτέροις/ Turnebus,Schneider; ἀμ ... Ω,Aldus; ἀμβλέσι Furlanus.
δέ· καὶ εἰσὶν/ δὲ κ. εἰ., ὥστε Hill; δέ· καὶ εἰσὶν οἱ τοιοῦτοι σκληρότεροι conj.Stephanides.
(After these words the rest is missing in B, and part of the Περὶ ἱδρώτων follows. The next line is: ... παρα. ἀεί τε συνεχὲς ὂν καὶ μὴ πολλῶν

· 37 ·

ἐνυπαρχόντων. παρα seems to be the beginning of παραπλησίως in the Περὶ λίθων.)
κάτω/ ἄτοπον conj.Wimmer (cf. ἄτοπον δὲ κἀκεῖνο, §44).
κάτω τὰ . . . μὴ τέμνεσθαι . . ./ κατὰ τὸ μὴ τ. Furlanus; κάτω· τὰ δὲ μὴ τ. Turnebus; ἄτοπον τὸ ὅλως μὴ τέμνεσθαι λίθον conj.Stephanides.
σιδήρῳ/ Aldus (σιδήρῳ); . . . ρω Ω.
ἔτε . . ./ Ω; ἔτε Aldus,Wimmer.
στερεὸν ἔτε/ στερεώτερα καὶ Turnebus, Stephanides.
καὶ σίδηρος/ καὶ om.Turnebus.
λίθου/ Turnebus; λίθους Ω,Aldus.
44 ᾧ γλύφουσι/ Turnebus; τῶ γ. Ω,Aldus.
ἡ ⟨ἀρίστη⟩/ Schneider; ἡ Ω,Aldus.
45 τὴν αὐτὴν/ Schneider; τὴν τοιαύτην Ω,Aldus.
ἔχειν τῷ πυρὶ δύναμιν/ Aldus; δύναμιν ἐ. τ. π. Ω.
ἐκεῖνο/ Aldus; ἐκεῖ Ω.
τῷ τὰ/ Aldus (τῶ); τὸ τὰ Ω.
ἀλλοιοῦν/ Turnebus, Schneider; ἀξιοῦν Ω; ἰοῦν Eichholz.
46 κατάχαλκον χρυσὸν/ Schneider; χαλκὸν κατὰ χρυσὸν A,Aldus; χαλκὸν καταχρυσὸν C; χαλκὸν κατάχρυσον Turnebus.
κριθή/ Turnebus; κριθήν Ω,Aldus.
κόλλυβος/ Schneider; κόλιμβον A; κόλυμβον C,Aldus; κόλλυμβος Turnebus; κόλυβον Furlanus.
ἡμιώβολος/ Schneider; ἡμιόλιον ὀβολὸν Ω; Aldus (ὀβ-); ἡμιοβόλιον ὀβολὸν Turnebus; ἡμιόβολος Furlanus.
47 πᾶσαι ἐν τῷ ποταμῷ/ ἐ. τ. π. πᾶσαι, corr. above line A.
τὰ ἄνω/ τὰ ἄνω (τὰ) Schneider.
δὲ διότι/ Wimmer; δέον· ὅτι Ω,Aldus; δέον, ὅτι Turnebus; δέ, ὅτι Schneider (Syll.) from Coraës.
δοκιμάζει χεῖρον/ Turnebus,Schneider; τοῦ δοκιμάζειν χ. Ω,Aldus; τὸ δοκιμάζειν χ. Furlanus.
ὑπολαμβάνουσιν/ Schneider; ὑπολαμβάνει ὡς Ω,Aldus; ὑπολαμβάνουσιν, ὡς Turnebus.
τι τοῦ ἔδους/ Schneider (Syll.) from Coraës; τὸ τοῦ εἴδους Ω,Aldus.
48 εἰσὶν/ Schneider; σχεδόν εἰσιν Ω,Aldus.
μαλάττεσθαι/ Turnebus, Schneider; ἀλλοιοῦσθαι Ω,Aldus.
τοῖς χυτοῖς καὶ ὀρυκτοῖς/ ὁμοίως τοῖς ὀρυκτοῖς καὶ χυτοῖς conj.Schneider.
λίθους/ πλίνθους Furlanus.
ποικίλας/ ποικιλίας Turnebus, Schneider.
συντιθεμένας . . ./ τὰς σ. Ω,Aldus; τὰς συνθέσεις Turnebus; συνθέσεις Schneider; συντιθεμένας ἐκ τῆς 'Ασίας ἄγουσιν conj.Stephanides (cf. §61).
ἁπάσας γὰρ/ γὰρ om.Turnebus, Schneider.
49 ὑελίτιδος/ Hill; ὑελίδος Ω,Aldus.
ὥς τινες/ οὕ τινες A.
πυκνώσει/ πυρώσει Hill.

APPARATUS CRITICUS

χαλκῷ/ χάλικι Hill (conj.Laet).
τῷ τήκεσθαι/ Turnebus, Schneider; τὸ τ. Ω,Aldus.
ἔχει/ C,Turnebus; ἔχειν A,Aldus.
50 εἴη/ Turnebus; ἢ A, ἢ C, ᾖ Aldus.
τῶν τόπων/ Wimmer; τοὺς τούτων Ω,Aldus; τοὺς τῶν τόπων conj. Schneider.
⟨ἰδίαν⟩ τιν'/ Schneider; τιν' Ω,Aldus.
ἔχουσι/ Turnebus,Schneider; ἔχουσαι Ω,Aldus.
τοὺς τῶν φυτῶν/ τὰς τ. φ. Furlanus,Schneider.
⟨αὐ⟩τὰς/ Wimmer; τὰς Ω,Aldus; τοὺς Furlanus; αὐτὰς Schneider.
διαριθμήσειεν/ διαριθμήσεις Aldus; διαριθμήσειε Furlanus.
51 ἀργυρείοις/ ἀργυρίοις A.
τε καὶ/ τε om.Aldus; ⟨τε⟩ καὶ Furlanus.
ἔνια/ ἤια A.
φασιν/ Heinsius; φαντασίαν Ω,Aldus.
γραφεῖς/ Furlanus; βαφοῖς A; βαφεῖς C,Aldus.
ἀρρενικοῦ/ Aldus; ἐρρενικοῦ Ω.
52 ἀλλὰ/ Furlanus; τᾶλλα A; τὰ ἄλλα C; τἄλλα Aldus.
μεταλλεῦσι/ Turnebus,Schneider; μετάλλοις Ω,Aldus.
δοκεῖ/ Turnebus; ποιεῖ Ω,Aldus.
σιδηρεῖα/ Schneider; σιδήρια Ω,Aldus.
Λημνία/ Heinsius (λ.); λιμνία Ω,Aldus.
καὶ ἦν/ Furlanus; καὶ om.Ω,Aldus.
καλοῦσι Σινωπικήν/ Schneider; καλοῦσιν ὠπτικήν Ω,Aldus; κ. συνωπικήν Turnebus; κ. Σινωπικήν Furlanus.
53 τῷ μικρῷ/ Ω,Aldus; τῇ Λήμνῳ Hill, conj.Furlanus.
μεταλλεύεται/ μεταλεύεται C.
γένη/ γεννή A.
μιγνύουσι/ Turnebus; μισθοῦσι Ω,Aldus.
δὲ καὶ ἐκ/ Schneider; τ' ἐκ Ω,Aldus; δὲ ἐκ Furlanus.
ὤχρας/ Aldus; χώρας Ω.
πανδοχείου/ Furlanus; παντωλίου Ω,Aldus; παντοπωλίου Turnebus.
54 καινὰς/ Turnebus, Schneider; κενὰς Ω.
πήλῳ/ Turnebus; πολὺν Ω(?A),Aldus.
ὀπτῶσι/ Aldus; ὀπτῶσαι Ω.
γενόμεναι/ Schneider; γινόμεναι Ω,Aldus.
μαρτυρεῖ δ'/ Schneider; μ. δ' ἂν Ω,Aldus.
αὐτή/ Schneider; αὐτό Ω,Aldus.
δόξειε γὰρ ἄν/ Turnebus, Schneider; δ. γ. τὸ Ω,Aldus.
μεταβαλεῖν/ Schneider; μεταβάλλειν Ω,Aldus.
εἴπερ/ ? ἤπερ C.
τῇ φυσικῇ/ τ. φ. ⟨γενέσει⟩ Schneider.
55 χυτὸν/ τεχνητὸν Furlanus.
φόρον/ Turnebus; φόρου Ω,Aldus.
τὸν μὲν κύανον/ τὸν Σκύθην κ. Eichholz.

· 39 ·

λευκότατον/ Turnebus; λεπτότατον Ω,Wimmer.
τῶν παχυτάτων/ Ω,Aldus; τῶν om.Turnebus,Wimmer.
ψιμύθιον/ Heinsius; ψιμίθιον Ω,Aldus; also in §56.
56 μόλυβδος/ Wimmer; μόλιβδος Ω,Aldus.
ἡλίκον πλίνθος (after πίθοις)/ Schneider; ἡλίκον πλῆθος (after πάχος) Ω,Aldus.
πάλιν/ Aldus,Turnebus; πάλιν τιθέασι καὶ πάλιν Ω.
ἀφηθοῦσιν/ Schneider; ἀφιθοῦσιν Ω; ἔφθουσιν Aldus.
57 τρυγὸς/ Heinsius; τριγὸς Ω,Aldus.
ἐπιγινόμενον/ C,Hill; ἐπιγενόμενον A,Aldus.
αὐτῷ/Wimmer; οὕτω Ω,Aldus.
γὰρ ὁ ἰός/ Wimmer; τιθέμενος Ω,Aldus.
αὐτῷ· ἐπιφαίνεται γὰρ ὁ ἰός/ om.Schneider; οὕτω δὲ ἐπιφαίνεται ὁ ἰὸς τιθεμένου (after τίθεται) Schneider (Syll.).
58 ἐργασίαν/ Aldus; ἐργασίας Ω.
⟨ἐπὶ⟩ κρημνῶν/ Schneider; κρημνῶν Ω,Aldus.
ὃ καταβάλλουσι/ Aldus,Turnebus; ὃν κ. Ω.
μόνον. ἔστι δ'/ Schneider; μ. δ' ἐστὶν Ω,Aldus.
λαμπυρίζουσαν/ Aldus; λαμπυρίζουσι Ω.
[μικρὸν ἐν καλοῖς]/ Schneider; μ. ἐ. κ. Ω,Aldus. A repetition of ἐν χαλκοῖς, conj.Wimmer; ἐν ἀγροῖς Κιλβιανοῖς (after ὑπὲρ Ἐφέσου μικρὸν above), conj.Schneider.
οἱ μὲν γὰρ/ οἱ μὲν καὶ Aldus.
⟨τῷ⟩ ἐπάνω/ Schneider; ἐπάνω Ω,Aldus.
κιννάβαρι (after κάτω)/ κινάβαρι A.
59 ἔχει/ ἔχοι A.
ἐνενήκοντα/ ἐννενήκοντα C.
60 μιμεῖται/ ⟨τὰ μὲν⟩ μιμεῖται Schneider.
τὰ δ' ἴδια/ τὰ δὲ ⟨καὶ⟩ ἴ. Schneider.
ἄλπεις/ ἀλιπεῖς Furlanus; perhaps ἀλιφάς (Suppl.Epigr.Gr.III,147) or ἀλοιφάς; χαλκομιγεῖς (sc. λίθους) conj.Stephanides (cf.§49).
χρεία/ Furlanus; χρόα Ω,Aldus.
τὸ ⟨κιννάβαρι⟩ τριφθῇ/ Schneider; τι . . . τριφθῇ Ω,Aldus; τι ⟨κιννάβαρι⟩ τριφθῇ Furlanus.
δοίδυκι/ δύδυκι A.
61 γεωφανέσιν/ Ω,Aldus,Schneider; γεωφανέσι Wimmer.
[περὶ] ὧν/ Schneider; περὶ ὧν A (? περὶ), C,Aldus.
ὑποκειμένων . . ./ ὑ. καὶ Turnebus (also conj.Schneider); ὑ. τε καὶ Laet.
. . . ουντων/ ποιούντων Turnebus; διηθούντων conj.Schneider.
μαλάττοντες/ Schneider; μελαντῶντες Ω,Aldus.
ταύτας/ εἰς ταύτας Laet.
62 Μηλιὰς/ Laet (μ.); μιλιὰς Ω,Aldus; μυλιὰς Turnebus; Μηλία Heinsius.
Τυμφαϊκὴ/ Turnebus (τ.); στυμφαϊκη Ω,Aldus.
τετάρτη/ Hill; καὶ τ. Ω,Aldus.
παρὰ ταύτας/ Turnebus; περὶ τ. Ω,Aldus.

APPARATUS CRITICUS

ἢ γύψος/ Hill; ἡ γ. Ω,Aldus.
Μηλιάδι/ Laet (μ.); μιλία διὰ Ω; μιλίᾳ διὰ Aldus; μυλίᾳ διὰ Turnebus; μιλίᾳ δὴ Furlanus; Μηλίᾳ δὴ Heinsius.
τῇ Σαμίᾳ/ Turnebus (σ.); τῆς μιᾶς Ω,Aldus.
ἤρεμον/ ἀραιὸν Turnebus; ἀραιὸν ἤρεμον Furlanus.
καὶ . . . δες/ καὶ τραχῶδες Turnebus, Furlanus.
ἀλιπὲς/ ἐλλιπὲς C.
Μηλιὰς/ Laet (μ.); μιλιὰς Ω,Aldus; μυλιὰς Turnebus.
τῷ φαρίδι/ τῷ φαριδ' (no accent) Ω; τῷ Φάριδι Aldus; ἐν τῷ Φάριδι Heinsius. ἐν τῷ ψαφαρῷ or σὺν τῇ ψαφαρότητι conj.Schneider. Possibly διάφορον.
63 εἰσὶ δὲ/ Laet, Schneider; ἔστι καὶ Ω; εἰσὶ καὶ Aldus.
στῆναι/ Schneider; στῆσαι Ω,Aldus.
τὸ μὲν/ Aldus; τὰ μὲν Ω.
ἡλίκη/ Furlanus; ἡλικὴν Ω,Aldus.
δίπους/ Turnebus; διπλοῦς Ω,Aldus.
περιέχουσιν/ Furlanus; περιέξουσιν Ω,Aldus.
διαφυὴν/ Furlanus; διαφύειν Ω,Aldus; perhaps διαφυὴν ⟨δ'⟩.
βελτίων/ Furlanus; βέλτιον Ω,Aldus.
αὐτῆς/ Furlanus; αὐτὴ A; αὐτῇ C; αὐτῇ Aldus; om.Laet.
ἄχρι/ Turnebus; . . . ταχρη Ω,Aldus.
τεττάρων . . . ἐστὶν/ Schneider; τεττάρων ἐστὶν Ω,Aldus.
ἐστὶν ἡ ἐσχάτη/ ἐστὶν, ἡ ἐ. Turnebus; ἐστίν. ἡ ἐ. Furlanus; ἐστὶν ἔχουσα. ἡ ἐ. Laet.
64 μάλιστα ἢ/ μάλλιστ' ἢ A, μάλιστ' ἢ C.
μάλιστα . . . ἱμάτια/ Bracketed by Furlanus.
ἢ μόνον/ Κιμωλίᾳ Hill.
χρῶνται δὲ καὶ/ χ. δὲ Ω,Aldus; χ. δὲ ⟨καὶ⟩ Schneider.
Τυμφαϊκῇ/ Turnebus (τ.); τυφικῆς Ω; τυφικῇ Aldus.
τὸν Ἄθων/ Schneider; θεάτων Ω,Aldus; τε Ἄθων Turnebus; Τυμφαίαν Heinsius.
καὶ τοὺς/ Turnebus; δὲ καὶ τοὺς Ω,Aldus.
γύψος/ γίψος A.
Τυμφαίαν/ Furlanus; τυμφετὸν Ω; τύμφεται Aldus; τύμφας τε Turnebus.
Περραιβίαν/ Schneider; περεβίαν Ω; περαιβίαν Aldus.
65 αὐτῆς/ Schneider; αὐτῶν Ω,Aldus.
ἀλαβαστρίτῃ/ Aldus; ἀλαβάστριτις Ω.
χαλικώδης/ Furlanus; χαλκώδης Ω,Aldus; πλακώδης Turnebus.
βρεχθῇ/ Turnebus; ἐρεχθῇ Ω,Aldus; ἐρεχθῇ Furlanus.
τὸν λίθον/ τοῦτον τ. λ. Ω,Aldus; [τοῦτον] τ. λ. Schneider.
περιχέοντες/ Schneider; περιέχοντες Ω,Aldus.
τοιοῦτο/ Ω,Aldus; τοιοῦτον Schneider,Wimmer.
66 ἐὰν ⟨δὲ⟩/ Schneider; ἐὰν Ω,Aldus.
διελεῖν/ Schneider; διελθεῖν Ω,Aldus.
θαυμαστὴ δὲ καὶ ⟨ἡ⟩ ἰσχύς/ Schneider (Syll.); θαῦμα ἐστι δὲ καὶ ἰσχύς

· 41 ·

THEOPHRASTUS ON STONES

Ω; θαῦμα. ἐστὶ δ. κ. ἰ. Aldus,Turnebus (ἔστι); ἅμα. ἔστι δ. κ. ἰ. Furlanus; θαυμασία δ. κ. ⟨ἡ⟩ ἰ. Schneider.
ὅτε γὰρ/ Ω,Schneider (Syll.); ὅτε τε Aldus; ὅτι Turnebus; ὅτε Schneider.
λίθοι/ τοῖχοι Furlanus.
ἢ διαφέρονται/ Schneider (Syll.); καὶ διαφέρονται Ω,Aldus,Turnebus, Schneider; καὶ διαφθείρονται Furlanus.
ἡ γύψος οὐκ/ ἡ δὲ γ. οὐκ Schneider (Syll.); ἡ δὲ ... μος Ω (ο ... μος A), Aldus,Schneider; ἢ δεσμὸς Turnebus; ἡ δὲ ἄμμος Furlanus.
μένει συνεχόμενα/ Schneider (Syll.); καὶ συνεχόμενα Ω,Aldus.
ὅτε γὰρ ... κολλήσει/ ὅτε οἱ λίθοι ῥήγνυνται ἢ διαφέρονται, ἡ δὲ γύψος οὐκ ἀνίησι· πολλάκις δὲ καὶ τὰ μὲν πέπτωκε καὶ ὑφῄρηται, τὰ δ' ἄνω κρεμάμενα μένει συνεχόμενα τῇ κολλήσει Coraës, quoted by Schneider (Syll.).
67 ὑφαιρουμένη/ Furlanus; ὑφαιρομένη Ω,Aldus.
καὶ πάλιν/ Turnebus; καὶ πα ... Ω,Aldus.
μάλιστα/ μάλιστα ⟨χρῶνται⟩ Schneider.
τὸν οἶνον/ Turnebus, Schneider; τὸν οἰκεῖον Ω,Aldus; τὴν κονίασιν Hill (conj.Salmasius; cf. Schneider IV, p. 592).
⟨εἰς⟩ ἔνια/ Wimmer(D); ἔνια Ω,Aldus; ⟨πρὸς⟩ ἔνια Schneider,Wimmer (T).
γναφεῖς/ Wimmer; κναφεῖς Ω,Aldus.
τὰ ἀπομάγματα/ Ω,Aldus; τὰ om.Schneider,Wimmer.
καὶ μάλισθ'/ ἢ μ. Ω.
γλισχρότητι/ Turnebus; ... χρότητι A,Aldus; ὠχρότητι C; ⟨διὰ τὴν⟩ γλισχρότητα Schneider.
λειότητι/ λειότητα Schneider.
68 ἡ μὲν δύναμις/ ἡ μὲν ⟨οὖν⟩ δ. Schneider.
τοῖς τοιούτοις/ Ω,Aldus; τοιούτοις Schneider,Wimmer.
τὰ τῆς κονίας/ Schneider; κατὰ τῆς κ. Ω,Aldus; κατὰ τὰ τῆς κ. Laet.
τὰ τῆς γῆς/ A, Aldus; κατὰ τῆς γ. C; κατὰ τὰ τῆς γ. Laet.
ἑκατέραν ὑπερέχουσαν/ Schneider; ἑκατέρας ὑπερεχούσας Ω,Aldus.
After ὑπερέχουσαν/ θερμοτέρα γὰρ τῆς κονίας, γλισχροτέρα δὲ πολὺ τῆς γῆς add.Ω,Aldus; om.Schneider,Wimmer.
ἔμπυρος/ Aldus; ἔμπειρος Ω.
ἱματηγὸς/ Turnebus; ἱματιγὸς Ω,Aldus.
αὐτή/ Laet; αὕτη Ω,Aldus.
69 [καὶ καίοντες]/ Schneider; καὶ καίοντες Ω,Aldus.
καὶ ἁπλουστέρους, στερεωτάτους μὲν παρατιθέντες/ καὶ ἁπλῶς τοὺς στερεωτάτους, παρατιθέντες conj.Schneider.
⟨βόλιτον, ἕνεκα⟩ τοῦ/ Schneider; ... τὰ τοῦ Ω,Aldus, ἕνεκα τοῦ Turnebus; Eichholz; διὰ τὸ Furlanus.
μὲν παρατιθέντες/ βόλιτον π. Eichholz.
τούτου/ Aldus; τούτων Ω.
ὅτι/ Furlanus; ὅτι τὸ Ω,Aldus.
αὐτῆς/ Schneider; αὐτὴ Ω,Aldus.

· 42 ·

TRANSLATION

TRANSLATION

ON STONES

Of the substances formed in the ground, some are made of 1
water and some of earth. The metals obtained by mining, such as
silver, gold, and so on, come from water; from earth come stones,
including the more precious kinds, and also the types of earth
that are unusual because of their color, smoothness, density, or any
other quality. As the metals have been discussed in another place,
let us now speak about the stones.

In general we must consider that all of them are formed from 2
some pure and homogeneous matter as a result of a conflux or
percolation, or because the matter has been separated in some other
way, as has been explained above. For perhaps some are produced
in one of these ways, and some in the other way, and others in a
different manner. Hence they gain their smoothness, density,
brightness, transparency, and other such qualities, and the more
uniform and pure each of them is, the more do these qualities
appear. In general, the qualities are produced according to the accuracy with which the stones are formed and solidified.

Some things are solidified through heat, others through cold. 3
And probably there is nothing to prevent some kinds of stones
being formed by either of these two methods, although it would
seem that all the types of earth are produced by fire, since things
become solid or melt as a result of opposite forces. There are more
peculiarities in stones; for most of the differences in the types of
earth concern color, tenacity, smoothness, density, and so on, but
in other respects the differences are rare.

Stones, however, have these differences and in addition there 4
are others that depend on their power of acting on other substances, or of being subject or not subject to such action. For some
can be melted and others cannot, some can be burnt and others
cannot, and there are other differences of this kind. And some
show a number of differences in the actual process of being set
on fire and burnt, and some, like the *smaragdos*,[1] can make the

---

[1] A green stone, mentioned again in secs. 8, 23, 24, 25, 26, 27, 35. The usual English translation is "emerald," but this is not the exact meaning of the word. For the identification of this stone, see Commentary, secs. 23 and 24.

color of water the same as their own, whereas others can turn what is placed on them entirely into stone; some have the power of attraction and others can test gold and silver, such as the stone called the Heraclean and the one called the Lydian.

5 But the greatest and most wonderful power, if this is true, is that of stones which give birth to young. But the power of those used in manual work is better known and is found in more varieties. For some can be carved, or turned on a lathe, or sawn; there are some on which an iron tool cannot operate at all, and others on which it works badly and with difficulty. And there are several other differences in addition to these.

6 The differences that are due to color, hardness, softness, smoothness, and other such qualities, through which stones gain their special excellence,[2] are found in many varieties, and in some they occur in the whole of a district. And among such stones there are the Parian, the Pentelic, the Chian, and the Theban, and these stone quarries have become widely known. There is also the *alabastrites*,[3] found at Thebes in Egypt—this, too, can be worked in large blocks—and the stone resembling ivory which is called *chernites*;[4] and they say that Darius was buried in a sarcophagus of this

7 material. And there is the ⟨variety of⟩ *poros*,[5] which is like Parian marble in color and density, but has only the lightness of ⟨ordinary⟩ *poros*; for this reason the Egyptians use it as a frieze in their elaborate buildings. And a dark stone is also found in the same place, which is translucent like the Chian stone, and there are several other kinds in other places. Such differences are common to many stones, as we have already said, but those that are due to the powers mentioned above are not found now in whole districts or in continuous or large masses of stone.

8 Some stones are quite rare and small, such as the *smaragdos*, the *sardion*,[6] the *anthrax*,[7] and the *sappheiros*,[8] and almost all those

[2] See Commentary.   [3] In all probability, onyx marble.
[4] Apparently a variety of onyx marble.
[5] This could be travertine, but here it is probably a special kind of *poros* found in Egypt.
[6] A red stone mentioned again in sec. 23 and described briefly in sec. 30. See Commentary, sec. 30, for its identification.
[7] Another red stone which is described in secs. 18 and 19. See Commentary, secs. 18 and 19.
[8] A blue stone which is described briefly in secs. 23 and 37. For its identification, see Commentary, sec. 37.

TRANSLATION

that can reasonably be cut and used as seals.⁹ And some are discovered in other stones when these are cut up. There are a few which can be set on fire and burnt, and perhaps we should first explain the nature and extent of their differences.

Some of them melt and become fluid when subjected to fire, such as those which come from mines. For when silver, copper, and iron become fluid, so does their stony matrix, either because of the moisture in the matter it contains or because of the nature of such stones. In this way, too, fire-resisting stones and millstones become fluid along with the material placed on them by those who are burning it. And some go so far as to say that all of them melt except marble and that this burns up and lime is formed from it. 9

But it would seem that it is going too far to say this; for there are many which break and fly into pieces as if they are fighting against being burnt, like pottery, for example. And this is natural since they have lost their moisture; for whatever can be melted must be moist and have a good deal of humidity. 10

And they say that some stones that are exposed to the sun become completely dry, so that they are useless unless they are soaked and wetted again, and others become softer and are more easily broken. It is clear that both kinds are robbed of their moisture by the sun, but it happens that stones of solid texture become hard when they are dry, whereas those that are loose in texture, and those whose formation is of this kind, are easily broken and melted. 11

Some of those that can be broken are like hot coals when they burn, and remain like this for some time, such as those found in the mine at Binai¹⁰ which are brought down by the river; for when they are covered with charcoal they burn as long as air is blown onto them, then they die down and afterwards can be kindled again, so that they can be used for a long time, but their odor is very harsh and disagreeable. 12

There is a stone called *spinos*,¹¹ which was found in mines. If this is cut up and the pieces are piled in a heap, it burns when 13

⁹ The emendation ⟨πάντες τῶν κατὰ⟩ has been added to the text.
¹⁰ The text uses the plural form (Binai), but the place was usually known as Bina.
¹¹ Probably some sort of asphaltic bitumen.

exposed to the sun, and it does this all the more if it is moistened and sprinkled with water.

14 But the Liparean stone[12] is made porous when it is burnt, and becomes like pumice, so that both its color and density are altered; for before it has been burnt it is black, smooth, and compact. This stone is found in pumice, appearing separately in various places and not continuously, as if it were in a cell of a honeycomb. In the same way it is said that in Melos pumice is found in another kind of stone, and so the Liparean stone corresponds to this in the opposite way, as it were, except that this stone is not the same as the Liparean stone.

15 The stone which is found at Tetras in Sicily also becomes porous. This place is in the neighborhood of Lipara, and the stone is plentiful in the promontory called Erineas. Like the stone found at Binai, it releases a bituminous odor when it is burnt, and what remains after the burning is similar to burnt earth.

16 Among the substances that are dug up because they are useful, those known simply as coals are made of earth, and they are set on fire and burnt like charcoal. They are found in Liguria, where amber also occurs, and in Elis as one goes by the mountain road to Olympia; and they are actually used by workers in metals.

17 In the mines at Scaptē Hylē a stone was once found which was like rotten wood in appearance. Whenever oil was poured on it, it burnt, but when the oil had been used up, the stone stopped burning, as if it were itself unaffected. These are roughly the differences in the stones that burn.

18 But there is another kind of stone which seems to be of an exactly opposite nature, since it cannot be burnt. It is called *anthrax*, and seals are cut from it; it is red in color, and when it is held towards the sun it has the color of a burning coal. One might say that it has great value; for a very small one costs forty pieces of gold. It is brought from Carthage and Massalia.

19 The stone found near Miletus does not burn; it is angular and there are hexagonal shapes on it. It is also called *anthrax*, and this is remarkable, for in a way the nature of *adamas*[13] is similar.

This power of resisting fire does not seem to be due to the

[12] Obsidian.
[13] Probably corundum. See Commentary, secs. 19 and 44.

absence of moisture, as is true of pumice and ashes. For these cannot be set on fire and burnt, because the moisture has been removed, and some think that pumice is formed entirely as a result of burning, with the exception of the kind that is produced from the foam of the sea. Their belief is due to observation and is 20 based on what is produced in craters of volcanoes and also on the porous stone[14] which changes[15] to pumice when it is fired. And the places where it is produced seem to prove this, for pumice is found especially in places that ....[16]

But perhaps one kind is made in this way, and others in another 21 way, and there are many methods of producing it; for the pumice found in Nisyros seems to consist of a kind of sand. And it is regarded as proof of this that some of the stones which are found break into pieces in one's hands and crumble into sand, as it were, because they have not yet become compact and solid. People find them in groups but in small amounts,[17] mostly about a handful in size or a little larger, whenever they scrape off the surface covering them. And the sand is very light. The kind found in Melos is all ... ,[18] but some are produced in a stone of a different sort, as has been mentioned before.

They differ from one another in color, density, and weight. 22 They differ in color because the kind that comes from the lava stream in Sicily is black, and this stone and the *malodes*[19] differ in density and weight; for a pumice of this kind, having both weight and density, is also produced, and this is more valuable than the other in its practical use. The one that comes from the lava stream can cut better than the white kind, which is light in weight, but the kind that comes from the sea itself cuts best of all. So much for pumice. But we must consider elsewhere the causes

---

[14] *diabaros*, perhaps *diaboros*.
[15] The Aldine text has οὐ (i.e., "which does not change"). Wimmer accepts this, but ἤ is found in the manuscripts (i.e., "which changes").
[16] The emendation καιομένοις would mean "places that are burning"; πυρικαύστοις would mean "places that have been subjected to burning."
[17] κατὰ μικρά is used sometimes by Aristotle in this sense. Cf. *Meteorologica*, 370B, 5.
[18] The emendation σχεδὸν ὡς ἐν Νισύρῳ would mean "almost like the kind in Nisyros," but εὔθραυστος, ἐν λίθῳ δέ ... would mean "easily broken, but is produced in a stone."
[19] *Malodes* is unknown; *mēlodes* would suggest a pale-yellow stone, and *mylodes* would be a millstone.

THEOPHRASTUS ON STONES

of the difference between stones that either burn or do not burn, from which we moved into this discussion.

23 There are also other stones from which seals are cut that are ⟨remarkable⟩, some of them only for their appearance,[20] such as the *sardion*, the *iaspis*,[21] and the *sappheiros*, and the last of these seems to be spotted with gold. But the *smaragdos* also has certain powers, for it makes the color of water just like its own, as we have said before; a stone of moderate size affects a small amount of the water in which it is placed, the largest kind the whole of
24 the water, and the worst kind only the part close to it. It is also good for the eyes, and for this reason people carry seals made of it, so as to see better.[22] But it is rare and of small size, unless we are to believe the records about the Egyptian kings; for it is said that among the gifts from the king of the Babylonians a *smaragdos* was once sent to them which was six feet[23] in length and four and a half in width, and that four such stones are deposited as an offering in the obelisk of Zeus. These were sixty feet long, and their width was six feet at one end and three at the other. But these statements depend entirely on their writings.

25 The largest of the stones which many call *tanoi*[24] is the one at Tyre. For there is a large slab in the temple of Herakles, unless this is a false *smaragdos*, for a species of that kind does exist. The stone occurs in places that are well known and easy to reach, especially in two of them, the copper mines of Cyprus and the island lying off Chalcedon. In the latter, exceptional stones are found. This kind is obtained by mining, like the others, and nature has produced it separately in many veins in Cyprus.

26 They are not often found large enough for a seal, but most of them are smaller in size; for this reason the stone is used for soldering gold, since it solders like *chrysokolla*.[25] And some people even suppose that its nature is the same, for they both happen

---

[20] Schneider thinks that something is missing here; the meaning should be "some of them differ in their appearance but have the same name." Cf. sec. 30.
[21] Not our modern jasper. For its identification, see Commentary, sec. 27.
[22] Perhaps εὖ βλέπειν ("to see well," "to improve their sight").
[23] A cubit was about one and a half feet; thus four cubits means six feet.
[24] For a possible identification, see Commentary. Wimmer reads βακτριανῶν ("Bactrian stones") instead of τανῶν.
[25] A name applied to certain green copper minerals. It probably included malachite as well as the modern chrysocolla.

· 50 ·

TRANSLATION

to be similar in color. But *chrysokolla* is found in large quantities in gold mines and even more in copper mines, as in the ones near the . . . districts.

But *smaragdos* is rare, as we have mentioned, for it seems to 27
be formed from *iaspis*. It is said that a stone was once found in Cyprus half of which was *smaragdos* and half *iaspis*, as if it had not yet been entirely changed from the watery state. It takes some work to make it shine, for in its natural condition it is not bright.

It is remarkable in its powers, and so is the *lyngourion*;[26] for 28
seals are cut from this too, and it is very hard, like real stone. It has the power of attraction, just as amber has, and some say that it not only attracts straws and bits of wood, but also copper and iron, if the pieces are thin, as Diokles used to explain. It is cold and very transparent, and it is better when it comes from wild animals rather than tame ones and from males rather than females; for there is a difference in their food, in the exercise they take or fail to take, and in general in the nature of their bodies, so that one is drier and the other more moist. Those who are experienced find the stone by digging it up; for when the animal makes water, it conceals this by heaping earth on top. This stone needs working even more than the other kind.

And since amber is also a stone—for the kind that is dug up 29
is found in Liguria—the power of attraction would belong to this too. The stone that attracts iron is the most remarkable and conspicuous example. This also is rare and occurs in few places. This stone too should be listed as having a similar power.

There are several other stones from which seals are cut, such 30
as the *hyaloeides*,[27] which reflects images and is also transparent, and the *anthrakion*[28] and the *omphax*.[29] There is also rock crystal, and the amethyst, and both of them are transparent; and these two and the *sardion* are found when certain rocks are cut through. And there are others, as has been mentioned before, which differ from one another, though they have the same name. For one type

---

[26] Lynx-urine stone. Probably a variety of amber.
[27] Glasslike stone. Its identification is uncertain, but some possibilities are discussed in the Commentary.
[28] Mentioned again in sec. 33. For possible identification, see Commentary, sec. 33.
[29] Apparently a green stone of some sort. The problem of its identification is discussed in the Commentary.

of *sardion*, which is translucent and of a redder color, is called the female, and the other, which is translucent and darker, is called
31 the male. And it is the same with the varieties of the *lyngourion*, for the female is more transparent and yellow than the other. Also, one kind of *kyanos*[30] is called male and the other female, and the male is the darker of the two. The *onychion*[31] is mixed in color, with white and dark alternating; the amethyst is wine-colored. The *achates*[32] is also a beautiful stone; it comes from the river Achates in Sicily and is sold at a high price.

32 In the gold mines at Lampsakos a wonderful stone was once found, from which a seal was cut after it had been taken up to Astyra,[33] and this was then sent to the King because of its unusual nature.

33 These stones are rare as well as beautiful, but those that come from Greece are of less value, such as the *anthrakion* from Orchomenos in Arcadia. This is darker than the stone from Chios, and mirrors are made from it. There is also the stone from Troezen, and this is variegated with purple and white. The Corinthian stone is also variegated with the same colors, except that it is paler.

34 In general there are many stones of this kind, but the remarkable ones are rare and come from a few places only, such as Carthage, the country around Massalia, Egypt near the First Cataract, Syene
35 near the city of Elephantine, and the region called Psepho.[34] In addition, the *smaragdos* and the *iaspis* are found in Cyprus.

The stones which are used for mosaics come from Bactriana near the desert. They are collected by horsemen who go out at the time of the etesian winds; for they are visible then, since the violence of the winds disturbs the sand. But they are small in size and not large.

36 Among choice stones there is also the one called the pearl; this is translucent by nature, and valuable necklaces are made from it. It is produced in an oyster, which is similar to the *pinna* [except

---

[30] A variety of lapis lazuli. This same word also denoted certain blue pigments. See Commentary, secs. 31 and 55.
[31] A broader term than onyx. See Commentary.
[32] A name given to certain variegated stones, including some varieties of agate.
[33] This translates Ἄστυρα instead of στιράν. If Pliny is right, the king was Alexander the Great.
[34] Possibly Psebo.

that it is smaller. In size the pearl is as big as a large fisheye],[85] and it is produced in India and certain islands in the Red Sea.[86] These are approximately the stones that are of unusual excellence.

But there are some others, such as the fossil ivory which is variegated with white and dark markings. And there is also the stone called *sappheiros*, which is dark and not very different from the male *kyanos*, and there is the *prasitis*,[87] which has the color of verdigris. And the *haimatitis*[88] also is of solid texture; it is dull in color, and in accordance with its name seems to be made of blood that has become firm and dry. The stone called *xanthe*[89] is another variety, not really yellow in color but rather of a whitish tint, a color which the Dorians call *xanthos*. 37

Coral, which is like a stone, is red in color and rounded like a root, and it grows in the sea. And in a way the petrified Indian reed is not very different in its nature from coral. But this is a subject for another inquiry. 38

There are also many varieties of stones which are obtained by mining. Some of these contain gold and silver at the same time, but only the silver can be seen clearly. They are rather heavy in weight and have a strong odor. There is also natural *kyanos*, which contains *chrysokolla*, and there is another stone which is like glowing coals in color; and these stones are heavy. 39

In general a great many unusual types of such stones are found in mines; some of them are of an earthy nature, such as yellow ochre and red ochre, and some are sandy, like *chrysokolla* and *kyanos*, and others are powdery, such as realgar and orpiment and others that are like them. One could mention a number of peculiarities in such stones. 40

Some stones also have the power of not submitting to treatment, as we have mentioned before; for example, they cannot be cut with iron tools, but only with other stones. In general there is a great difference in the methods of working the larger stones; for some can be sawn, others can be carved, as has been stated, and others turned on a lathe, like the Magnesian stone. This is un- 41

---

[85] The words in brackets come from Athenaeus (III, 93).
[86] Probably the present Persian Gulf.
[87] An opaque green stone. For possible identification, see Commentary.
[88] Probably red jasper.
[89] Probably yellow jasper.

THEOPHRASTUS ON STONES

usual in its appearance, and some people are astonished at its resemblance to silver, though it is not related to it at all.

42 And there are a larger number of stones which submit to every method of treatment. In Siphnos there is a stone of this kind which is dug up about three furlongs from the sea; it is round and has the qualities of a lump of earth, and because it is soft it can be turned on a lathe and carved. When it is heated in the fire and dipped in oil, it becomes very black and hard; and dishes for the table are made out of it.

43 All such stones submit to the power of iron, but some stones, as we have mentioned before, can be carved with other stones, but not with iron instruments. And others can be carved with iron, but only with rather blunt tools. And . . . are . . . . In much the same way . . . cannot be cut with iron, and yet iron, which is harder than stone, cuts substances that are . . . stronger.[40]

44 That seems peculiar also, because whetstone wears away iron, although iron can divide the whetstone and shape it but cannot do this to the kind of stone from which seals are made. And again, the stone with which seals are carved consists of the material from which whetstones are formed, or something like it. And the ⟨best⟩ whetstone comes from Armenia.

45 The nature of the stone which tests gold is remarkable, for it seems to have the same power as fire, which can test gold too. On that account some people are puzzled about this, but without good reason, for the stone does not test in the same way. Fire works by changing and altering the colors, and the stone works by friction, for it seems to have the power of picking out the essential nature of each metal.

46 They say that a much better stone has now been found than the one used before; for this not only detects purified gold, but also gold and silver that are alloyed with copper, and it shows how much is mixed in each stater. And indications are obtained from the smallest possible weight. The smallest is the *krithē*, and after

---

[40] If certain emendations were adopted, this difficult passage might be translated as follows: "And such stones are harder. In much the same way it is peculiar that a stone cannot be cut at all with iron, and yet iron too, which is harder than stone, cuts substances which are firmer and stronger." (καὶ εἰσὶν οἱ τοιοῦτοι σκληρότεροι. παραπλησίως δὲ ἄτοπον τὸ ὅλως μὴ τέμνεσθαι λίθον σιδήρῳ· καίτοι καὶ στερεώτερα καὶ ἰσχυρότερα τέμνει καὶ σίδηρος λίθου σκληρότερος ὤν.)

that there is the *kollybos,* and then the quarter-obol, or the half-obol;[41] and from these weights the precise proportion is determined.

All such stones are found in the river Tmolos. They are smooth in nature and like pebbles, flat and not round, and in size they are twice as big as the largest pebble. The top part, which has faced the sun, differs from the lower surface in its testing power and tests better than the other. This is because the upper surface is drier, for moisture prevents it from picking out the metal. Even in hot weather the stone does not test so well, for then it gives out moisture which causes slipping. This happens also to other stones, including those from which statues are made, and this is supposed to be a peculiarity of the statue.

Of such kinds are the special qualities and powers found in stones. Earth has fewer of these, though they are more peculiar; for it is also possible for earth to be melted and softened and then hardened again. It melts ⟨along with⟩[42] substances which are dug up and which can be liquefied, just as stone also does. It is softened, and stones are made from it. These include the variegated ones and other composite stones . . . ;[43] for all of these are made artificially when they are fired and softened. And if glass is also formed, as some say, from vitreous earth, this too is made by thickening.

The most unusual earth is the one mixed with copper; for in addition to melting and mixing, it also has the remarkable power of improving the beauty of the color. And in Cilicia there is a kind of earth which becomes sticky when it is boiled, and vines are smeared with this instead of birdlime to protect them from woodworms.

It would also be possible to determine the differences that are naturally adapted for causing earth to turn into stone; for those that are due to locality, which cause different kinds of savors, have their own peculiar nature, like those which affect the savors

47

48

49

50

---

[41] See Commentary for a discussion of the absolute and relative values of these weights. They were probably equivalent to .06, .09, .18, and .36 of a gram.

[42] Some word is understood here, such as ἅμα ("along with"); cf. ἅμα in sec. 9. Schneider suggested ὁμοίως ("in the same way as").

[43] If ἐκ τῆς 'Ασίας ἄγουσιν were read, the meaning would be that "they bring them from Asia."

of plants. But it would be best to list them according to their colors, which painters also use.

These substances, as we said in the beginning, are produced either by some conflux or by percolation. Moreover, some seem to have been set on fire and burnt, such as realgar and orpiment and others of the same kind. To put it plainly, all of these result from a dry and smoky exhalation.

51 They are all found in silver and gold mines, and some of them in copper mines, such as orpiment, realgar, *chrysokolla*, red ochre, yellow ochre, and *kyanos*, but the last of these is seldom found and only in small quantities, whereas there are veins of some of the others, and yellow ochre is said to be found in masses. But there are all kinds of red ochre, so that painters can use it for flesh-colored pigments. And yellow ochre can take the place of orpiment, since there is no real difference in their color, though there seems to be.

52 But in some places there are mines that even contain both red ochre and yellow ochre together, as for example in Cappadocia, and large quantities are dug up. But they say that the risk of suffocation is a serious matter for the miners, since this can happen to them quickly and takes a very short time.

The best red ochre seems to be that of Ceos; for there are several kinds. One of them comes from mines, since iron mines also contain red ochre. But there is also the Lemnian kind and the one called Sinopic; this is really Cappadocian red ochre, but it is 53 brought down to Sinope. It is dug up by itself in . . . .[44] And there are three varieties of it, one very red, one light-colored, and a third whose color is midway between the others. We call this a self-sufficient kind because it does not have to be mixed, whereas the others do.

It is also made by burning yellow ochre, but this is an inferior kind and is a discovery of Kydias; for it is said that he became aware of it when an inn burnt down, as he noticed that some yellow ochre was half-burnt and had become red in color.

54 New earthen vessels are covered with clay and placed in ovens; for when the vessels become red-hot, they heat the ochre, and as they become hotter in the fire, they make its color darker and more

---

[44] ἐν τῷ μικρῷ is difficult; ἐν τῇ Λήμνῳ ("in Lemnos") is a possible alternative.

like glowing charcoal. And its origin is itself a proof of this; for it would seem that all these substances change under the influence of fire, if it is right to consider that the red ochre made in this process is the same as the one made by nature or very similar to it.

Just as there is a natural and an artificial red ochre, so there is a native *kyanos* and a manufactured kind, such as the one in Egypt. There are three kinds of *kyanos*, the Egyptian, the Scythian, and the Cyprian. The Egyptian is the best for making pure pigments, the Scythian for those that are more dilute. The Egyptian variety is manufactured, and those who write the history of the kings of Egypt state which king it was who first made fused *kyanos* in imitation of the natural kind; and they add that *kyanos* was sent as tribute from Phoenicia and as gifts from other quarters, and some of it was natural and some had been produced by fire. Those who grind coloring materials say that *kyanos* itself makes four colors; the first is formed of the finest particles and is very pale,[45] and the second consists of the largest ones and is very dark. These are prepared artificially, and so is white lead.

Lead about the size of a brick is placed in jars over vinegar, and when this acquires a thick mass, which it generally does in ten days, then the jars are opened and a kind of mold is scraped off the lead, and this is done again until it is all used up. The part that is scraped off is ground in a mortar and decanted frequently, and what is finally left at the bottom is white lead.

Verdigris is made in much the same way. Red copper is placed over grape-residues[46] and the matter that collects on it is scraped off; for it is verdigris that appears there.

There is also a natural and a prepared kind of cinnabar. The cinnabar in Iberia, which is very hard and stony, is natural, and so is the kind found in Colchis. They say that this is found on cliffs and is brought down by arrows that are shot at it. The prepared kind comes from one place only, a little above Ephesos. It is a sand that shines brightly and resembles scarlet dye; this is collected and ground in stone vessels until it is as fine as possible; then it is washed in copper ones . . . ,[47] and what remains is taken

---

[45] This translates λευκότατον, which has been accepted as an emendation in the text.
[46] The literal translation of τρύξ would be "wine-lees," but see Commentary.
[47] The words ἐν καλοῖς (after μικρόν) may simply be a repetition of ἐν χαλκοῖς. Schneider thinks they belong above, after ὑπὲρ Ἐφέσου μικρόν ("a little above Ephesos") and should really be ἐν ἀγροῖς Κιλβιανοῖς ("in the Cilbian district").

and ground again and then washed. Skill is needed for this process; for some people make a great deal and others little or nothing out of an equal amount of sand. The washing is done from the top, and separate portions are wetted one after the other; what is left at the bottom is cinnabar, and the washings are what remains above in larger quantities.

59 They say that Kallias, an Athenian from the silver mines, discovered and demonstrated the method of preparation; for thinking that the sand contained gold because it shone brightly, he collected it and worked on it. But when he saw that it did not contain any gold, he admired the beauty of the sand because of its color and so discovered this method of preparation. This did not happen long ago, but about ninety years before Praxiboulos was archon at Athens.

60 It is clear from these facts that art imitates nature and creates its own peculiar products, some of them for use, and some only for show, such as paints,[48] and others for both purposes equally, such as quicksilver; for this has its use too. It is made when cinnabar mixed with vinegar is ground in a copper vessel with a pestle made of copper. And perhaps several other things of this kind could be discovered.

61 Among the substances obtained by mining there still remain those that are found in earth-pits; these are caused, as we said in the beginning, from some conflux and separation of matter which is purer and more uniform than that of the other kinds. And all sorts of colors are obtained from them owing to the differences of the matter they contain . . . ;[49] some of them are softened and others are ground and melted, and in this way the stones that are brought from Asia are constructed.

62 The natural kinds of earth, which are useful as well as superior in quality, are three or four in number, the Melian, the Kimolian, the Samian, and a fourth in addition to these, the Tymphaic or *gypsos*.[50] Painters use only the Melian kind; they do not use the

---

[48] This translates ἀλιφάς; ἄλπεις is unknown. The emendation ἀλιπεῖς (lit., "without fat") must refer to the type of earth mentioned in sec. 62; this was not greasy and was suitable for painting.

[49] If καὶ διὰ τὴν τῶν διηθούντων were read, the meaning would be "and of the matter that percolates."

[50] A much broader term than the English word *gypsum*. It included not only the

Samian, even though it is beautiful, because it is greasy, dense, and smooth. For the kind which is ... and ...,[51] and is not greasy is more suitable for painting, and the Melian kind has this quality ....[52]

In Melos and Samos there are additional differences in the earth. It is not possible to stand upright while digging in the pits of Samos, but a man has to lie on his back or his side. The vein stretches for a long way and is about two feet in height, though much greater in depth. It is surrounded on both sides by stones and is taken out from the space between them. It has a stratum right through the middle, and this is better than the parts on the outside; and then it has another stratum, and still another, up to four .... The innermost earth is called "the star." This earth is used mainly or solely for clothes.[53]

The Tymphaic earth is also used for clothes and is called *gypsos* by the people who live near Mt. Athos and those districts. *Gypsos* occurs in large quantities in Cyprus and can easily be seen; for only a little soil is removed when it is dug up. In Phoenicia and Syria it is made by burning stones, and this also happens in Thourioi; for a large amount is produced there. And thirdly, it occurs in Tymphaia and in Perrhaibia and in other places.

Its nature is peculiar; for it is more like stone than earth, and the stone resembles *alabastrites*. It is not cut out in a large mass but in small pieces. Its stickiness and heat, when it is wet, are remarkable; for it is used on buildings and is poured around the stone or anything else of this kind that one wishes to fasten.

After it has been pulverized and water has been poured on it, it is stirred with wooden sticks; for this cannot be done by hand

---

mineral now called gypsum but also the roasted mineral. Lime and lime mortar also appear to have been included under this term. For a discussion of the various identifications, see Commentary, secs. 62, 64-69.

[51] The literal meaning of ἤρεμον is "quiet," but what it means here is not certain. The emendations ἀραιόν and τραχῶδες mean "of loose texture" and "rough." The first of these is very appropriate, since it indicates low density and is the opposite of πυκνόν. Furlanus reads ἤρεμον in addition to ἀραιόν, but the adverb ἠρέμα ("gently," "slightly") would be better.

[52] The emendation ἐν τῷ Φάριδι assumes that this is the name of a place. Schneider suggests ἐν τῷ ψαφαρῷ or σὺν τῇ ψαφαρότητι, implying that the earth is "liable to crumble." If the adjective διάφορον is concealed here, it would mean that the earth has a certain quality "in a marked degree."

[53] For cleaning or whitening clothes. See sec. 67.

because of the heat. And it is wetted immediately before it is used; for if this is done a short time before, it quickly hardens and it is impossible to divide it. Its strength, too, is remarkable; for when the stones are broken or pulled apart, the *gypsos* does not become loose, and often part of a structure falls down and is taken away, while the part hanging up above remains there, held together by the binding force. And it can even be removed and calcined and made fit for use again and again.

67

In Cyprus and in Phoenicia it is used mainly for these purposes, but in Italy it is also used for treating wine. And painters employ it for some parts of their art, and so do fullers, who sprinkle it on clothes. It seems to be far superior to other earths for taking impressions, and is generally used for this purpose, especially in Greece, owing to its stickiness and smoothness.

68 Its powers are seen in these and similar uses, but its nature is such that it seems, as it were, to contain the qualities both of lime and of earth, namely, heat and stickiness, or rather each of these in a marked degree. It is also clear from the following example that it has a fiery nature; for once a ship loaded with clothes was itself burnt when the clothes became wet and caught fire.

69 *Gypsos* is also burnt in Phoenicia and in Syria, where it is fired in a furnace. Marbles especially are burnt, and also the more ordinary kinds of stones, while cow-manure is placed alongside the hardest ones[54] to make them burn better and more quickly. It seems to become extremely hot when it has been set on fire, and stays hot for a very long time. And when it has been calcined, it is pulverized like ashes. From this it seems clear that its nature is entirely due to fire.

---

[54] If καὶ ἁπλῶς τοὺς στερεωτάτους, παρατιθέντες βόλιτον were read, the meaning would be "and in general the hardest kinds of stones, while cow-manure is placed alongside them." Eichholz translates ἁπλῶς as "absolutely."

COMMENTARY

COMMENTARY

1. *The metals obtained by mining, such as silver, gold, and so on, come from water.*

In general, the ideas of Theophrastus on the origin and nature of mineral substances are based on those of his predecessors and teachers, Plato and Aristotle. His brief statement about the formation of metals from water appears to be taken directly from Plato,[1] who believed that there were two primary types of substances having the nature of water: one of these was the liquid kind of water, which included any material that normally existed as a flowing liquid; and the other was the fusible kind of water, which included substances, such as metals, that could be made to flow by the action of heat. Plato describes the nature and formation of gold as follows: "Of all the substances which we have ranked as fusible kinds of water, that which is densest is formed of the finest and most uniform particles. This is a unique kind, tinged with a glittering and yellow color, that most precious of possessions known as gold, which has filtered through rocks and there congealed . . . ."[2]

The ideas of Aristotle on the origin of metals are somewhat more complex. According to him,[3] the metals originated from the imprisonment of vaporous exhalations in the earth, particularly in stones, where they were congealed by some sort of drying process, and as a result a metallic substance was generated. Since it was supposed that this process was similar to the freezing of water, the metals were considered by analogy to be water, but only in a certain sense. Aristotle believed that metals consisted of matter which might have become water but could no longer do so. He did not consider them as originating from qualitative changes in actual water. On the whole, so far as we can judge from his meager statements here, the ideas of Theophrastus on the origin of metals were in somewhat closer agreement with the ideas of Plato than with those of his immediate predecessor.

The statements of Theophrastus in this first section of the treatise also indicate that he was more inclined to follow the theories

[1] *Timaeus*, 58D.   [2] *Timaeus*, 59B.   [3] *Meteorologica*, III, 6.

of Plato on the origin of non-metallic mineral substances than those of Aristotle. Plato believed[4] that ordinary stone was formed when, in a mixture of earth and water, the latter changed to a form of air and rose to its own region; in doing so it gave a thrust to the surrounding air, which reacted on the remaining earth and compacted it into stone. He also thought, however, that certain kinds of stones were composed of a combination of earth and water,[5] and this idea is accepted by Theophrastus as shown by his statements in sections 9 and 10. But according to Aristotle, infusible stones were formed by the dry type of exhalation acting on earth,[6] though certain fusible stones were, like the metals, more of the nature of water.[7]

1. *stones, including the more precious kinds.*
The distinction here, as shown by subsequent passages, is primarily between the stones which occur in large masses in the form of rock and the various minerals of limited size and distinctive properties, particularly those that were used for seals.

1. *the metals have been discussed in another place.*
Probably this refers to Theophrastus' lost work on mines[8] mentioned by Diogenes Laertius.[9]

2. *as has been explained above.*
Though this appears to be a reference to a preceding passage that has dropped out, the close and logical development of the discussion up to this point seems to leave no room for the introduction of any other statement. Possibly this refers to a discussion in the lost treatise on mines mentioned by Diogenes Laertius or to a passage in some other treatise that may have immediately preceded the present one in the books of Theophrastus as they were originally written.

From the rather vague statements in this section it would appear that Theophrastus was somewhat more advanced than his predecessors in his theories about the formation of mineral substances; these tended even to approach modern views on the formation of minerals by crystallization from magmatic or aqueous solutions.

[4] *Timaeus*, 60C. [5] *Timaeus*, 61. [6] *Meteorologica*, III, 6.
[7] *Meteorologica*, IV, 10. [8] Περὶ μετάλλων. [9] V, 44.

COMMENTARY

The more advanced nature of the views of Theophrastus is evident when his statements are compared with Plato's theories about the formation of stones and the reason for the differences in their physical properties. This is what Plato says: "But when earth is compressed by air into a mass that will not dissolve in water, it forms stone, of which the transparent sort made of uniform particles is fairer, whereas the opposite kind is coarser."[10]

3. *Some things are solidified through heat, others through cold.*

Theophrastus seems to follow the theories of Aristotle about the solidification of mineral substances. According to Aristotle,[11] anything capable of being solidified was either water or a mixture of water and earth, and the agent that brought this about was either heat or cold. Ordinary water and analogous substances like molten metals were solidified by cold. On the other hand, a salt solution, since it left a solid residue after being treated at an elevated temperature, was considered to be solidified by heat. Aristotle did not recognize, as we do, any radical difference in the two phenomena. Cold was supposed to act by driving off the heat, and the moisture of the liquid was believed to accompany the heat in the form of vapor. Heat was supposed to act directly in driving off the moisture and leaving the earthy or solid matter behind. In other words, solidification of substances by either agent was considered to be a drying process. Aristotle also showed that the solidification of some substances could occur in both ways. Certain mixtures of earth and water, such as common mud, were in this class, for either heat or cold could readily bring them to a solid condition.

3. *it would seem that all the types of earth are produced by fire, since things become solid or melt as a result of opposite forces.*

The formation of ashes when materials are set on fire and burnt, and the resemblance of these to natural earthy substances, may well have been the basis of the idea that all earthy substances owed their origin to fire. In subsequent passages (secs. 50, 54, 69) Theophrastus refers again to the part that fire plays in the forma-

[10] *Timaeus*, 60C.   [11] *Meteorologica*, IV, 6-12.

tion of earths, and he says more in a specific way on this subject than either Plato or Aristotle, though in his attempts to explain the process he follows closely the teachings of the latter. His statement that the solidification or melting of earths is the result of opposite forces appears to be based directly on the following theories of Aristotle: "Bodies solidified by the dry-hot are dissolved by water which is the moist-cold, whereas bodies solidified by cold are dissolved by fire, which is hot. . . . For the opposite of the dry-hot is the cold-moist and what the one solidified the other will dissolve, and so opposites will have opposite effects."[12] Theophrastus means, therefore, that since all earthy substances are dissolved or dispersed by water, which is the cold-moist, it necessarily follows that they must all be solidified by the opposite agent, which is fire, the dry-hot. A mechanism for the softening or dispersion of earths by water is also given by Aristotle.[13] He believed that the pores of an earthy substance were large enough to admit water particles, and that the entrance of these caused the breaking up of the solid material. Solidification came about again on the expulsion of the water particles by heat, and the earthy substance then resumed its original condition.

4. *some . . . can make the color of water the same as their own.*

The manuscripts and Aldus have ἐξομοιοῦνται followed by the participle δυνάμενοι, but the infinitive ἐξομοιοῦν is expected, i.e., "they are able to make the color the same." Turnebus felt the need of a main verb and changed the last syllable of ἐξομοιοῦνται to φαίνονται; Furlanus preferred λέγονται, and this was accepted by Wimmer. But it is not certain that a verb of this sort is necessary; it may be that the participle δυνάμενοι is parallel to the participle ἔχοντες that precedes it, and the editors have not found it necessary to change this to ἔχουσι. If this is so, ἔχοντες . . . , ἔνιοι δὲ . . . δυνάμενοι is similar to τηκτοὶ . . . οἱ δ᾿ ἄτηκτοι above. Here οἱ μέν is understood to correspond to οἱ δέ. Turnebus and Schneider both think that οἱ μέν should be added to the text before τηκτοί, but, though this makes the meaning very clear, it is not essential, since μέν is not always expressed. As λέγονται is not in

[12] *Meteorologica*, IV, 6, 383A-383B.   [13] *Meteorologica*, IV, 9, 385B.

COMMENTARY

the manuscripts and cannot be regarded as a certain emendation, it has been omitted and the last syllable of ἐξομοιοῦνται has been bracketed as a mistake.

4. *others can turn what is placed on them entirely into stone.*

The basis of this statement, which at first sight seems a mere invention, may be that a stony layer of calcium carbonate or silica is deposited by underground water on objects buried in the ground or in caves. Pliny[14] states on the authority of Mucianus that mirrors, body-scrapers, clothes, and shoes buried in sarcophagi made of a certain stone are petrified. As Bailey[15] has suggested, it is probable that these sarcophagi were made of limestone. Under suitable conditions, the passage of water containing dissolved carbon dioxide through a limestone sarcophagus would dissolve calcium carbonate from its walls and deposit it on the objects inside. In a sense, then, the apparent petrifaction of these objects could be attributed to the nature of the stone from which the vessel was made.

4. *some have the power of attraction and others can test gold and silver, such as the stone called the Heraclean and the one called the Lydian.*

In particular, the stones having the power of attraction were *lyngourion*, amber, and lodestone, which are briefly described in sections 28 and 29. Heraclean stone was a common early name for native magnetic iron oxide, or lodestone, and is so used, for example, by Plato.[16] The name is apparently derived from the locality where it was discovered or obtained, but it cannot now be determined with certainty whether this was the Heraclea in Pontus or in Lydia or elsewhere, though it seems very probable that it was somewhere in Asia Minor. Theophrastus apparently uses the name here to denote one kind of touchstone. Likewise, Pliny[17]

---

[14] XXXVI, 131.
[15] K.C. Bailey, *The Elder Pliny's Chapters on Chemical Subjects* (London, 1929-1932), Part II, p. 252.
[16] *Timaeus*, 80C; *Ion*, 533D.
[17] XXXIII, 126. The text reads, "*Auri argentique mentionem comitatur lapis quem coticulam appellant, quondam non solitus inveniri nisi in flumine Tmolo, ut auctor est Theophrastus, nunc vero passim; alii Heraclium, alii Lydium vocant. . . .*"

mentions that a stone of this kind was called by some the Heraclean, by others the Lydian, stone. It is very likely, however, that Pliny took his information from this treatise of Theophrastus. Since the term "Heraclean stone" was sometimes used to designate the lodestone, the term "Lydian stone," derived from Lydia in Asia Minor, where it was discovered or first used, was perhaps the more correct ancient name for the touchstone. Moreover, the term "Lydian stone" is the one that has come down into modern mineralogical literature. If, as stated by Hesychius,[18] the Heraclean stone derived its name from Heraclea in Lydia, this may account in part for the confusion of the two names in the works of ancient writers. On the other hand, the application of both these names to the same stone in the works of the later writers may have originated entirely from the somewhat ambiguous wording of this passage in Theophrastus. It seems almost certain that Theophrastus really intended to say here that "some have the power of attraction; others can test gold and silver, such as the so-called Heraclean stone and the Lydian stone *respectively*."[19] The properties and uses of the touchstone are described in sections 45, 46, and 47. See also the notes on these sections in the commentary.

5. *But the greatest and most wonderful power, if this is true, is that of stones which give birth to young.*
This curious idea seems to be the result of observing certain kinds of geode-like concretions that consist of an outer shell within which is contained a clayey, sandy, or stony nucleus. Sometimes the internal material is held so loosely that the concretion rattles when shaken. The ancients apparently believed that such stones were pregnant, and that the mineral matter on the inside was in the process of being generated. Bailey[20] thinks that the discovery of crystals with other smaller crystals attached was the origin of this idea. Though observations of such crystal growths may have contributed something to its origin, the available literary evidence gives more support to the other explanation.

[18] *S.v.* 'Ηρακλεία.
[19] Pliny's misunderstanding of this passage was pointed out by Salmasius. See *Plinianae Exercitationes* (1689), 776aF.
[20] *The Elder Pliny's Chapters on Chemical Subjects*, Part II, p. 253.

COMMENTARY

Although Theophrastus does not name specifically any stone having this power of procreation, other ancient writers give the name "eaglestone" to these concretions. Pliny[21] states that they were so named because they were found in eagles' nests, and the eagles were unable to hatch out their young without the aid of these stones. He distinguishes[22] four varieties of *aetites* or eaglestone according to the color or consistency of the shell or nucleus of the concretion, and he names the localities where the various kinds were found. It is evident from his descriptions that such stones were often clay ironstone concretions composed of siderite (native ferrous carbonate) or limonite (native hydrated ferric oxide) compacted with clay or sand. It is interesting to note that even in modern times the name "eaglestone" has been sometimes applied to such nodules of clay ironstone. However, if the accounts given by Pliny are accurate, the ancient term must have included other kinds of clayey or sandy concretions. He names in addition, and sometimes describes briefly, other stones which contained embryo stones within them, such as *cyitis*[23] and *gassinade*.[24] He even goes so far as to declare that the period of gestation for the second of these stones was three months. Possibly these were different concretions, though they may have been identical with *aetites*. As one might expect, eaglestones were worn in ancient times as amulets to prevent miscarriage.[25] Such stones have been worn for the same purpose in certain European countries in modern times.

Though Theophrastus appears to be serious in mentioning this supposed generative power of stones, his use of the phrase "if this is true" shows that he was actually skeptical about it. Such skepticism is much less evident in the statements of the other ancient writers who touch on this subject.

The word τικτόντων is an emendation adopted by Schneider and Wimmer instead of τικτῶν, the reading of the manuscripts. What is needed is the active participle of the verb τίκτω ("to bring forth young"). It seems almost certain that this verb is correct, since Theophrastus refers to a great and wonderful power and shows that he himself is doubtful about it. Furlanus and Heinsius

[21] X, 12, and XXXVI, 149.   [22] XXXVI, 149, 150, 151.
[23] XXXVII, 154.   [24] XXXVII, 163.
[25] Pliny, XXX, 130, and XXXVI, 151; Solinus, XXXVII, 14, 15.

miss the point by changing the word to τηκτῶν; for there would be nothing wonderful about substances that could be melted.

After γνωριμωτέρα δέ the word τῶν appears in the manuscripts, but it is rightly bracketed by Schneider and Wimmer, since a subject in the nominative is needed and not a genitive of comparison. This subject really requires the article to introduce the words κατὰ τὰς ἐργασίας; this has been understood in the translation of the passage ("the power of those used in manual work is better known"), and ⟨ἡ⟩ has been added to the text.

6. *through which stones gain their special excellence.* The manuscripts have δι' ὧν ("through which") before τὸ περιττόν, but Aldus and Turnebus changed this to διό ("wherefore"), and Furlanus suggested διά ("on account of"), which Wimmer accepted for his text. Schneider was doubtful about the phrase διὰ τὸ περιττόν. It is somewhat confusing, since it might mean "because of their unusual nature," but qualities like hardness and softness are not unusual. However, if τὸ περιττόν means that the stones have a special excellence which makes them valuable or unusual, then hardness might be appropriate as one such point of excellence. Theophrastus mentions εἴδη περιττότερα in section 1, and τὸ περιττόν may be a reference to this. If τὸ περιττόν is accepted here, there seems no reason why the manuscript reading δι' ὧν should be rejected in favor of the emendation of Furlanus; a relative clause meaning "through which they get their excellence" or "from which their excellence is derived" seems to be more appropriate here than a phrase meaning "on account of their excellence." There is no difficulty about understanding a verb for the relative clause, and δι' ὧν has been restored to the text.

6. *the Parian, the Pentelic, the Chian, and the Theban.* It is only in this section and the next one that Theophrastus mentions the marbles and other stones used for building or statuary. He seems to regard them merely as examples of stones in which certain distinctive properties can be seen throughout large masses.

The quarries on the island of Paros were celebrated in ancient times for their excellent marble. Though this is usually spoken of as Parian marble, it is also alluded to as the marble of Marpessos, from the particular mountain where the finest stone abounded.

COMMENTARY

Other places on the island give evidence of ancient operations, and the stone is still available for quarrying at the present day. From a dazzling white to a creamy white in color, it had such an attractive appearance that a number of ancient authors thought it worthy of mention. This was the marble which was regarded as most suitable for statuary and was used by many of the most celebrated sculptors of antiquity.

Scarcely less famous was the Pentelic marble, so named from Mt. Pentelicos near Athens. The ancient quarries were at a place called Spilia about twelve miles northwest of the Acropolis. This stone is still quarried at the present time, and most of it is used in the construction of buildings in modern Athens. In ancient times it was widely used for architectural purposes, as is shown by the Parthenon and other surviving structures. Pentelic marble is as fine-grained as marble from Paros, but it takes on a yellow tone on weathering, and there are occasional dark streaks running through it. It was apparently less esteemed than Parian marble for the purposes of sculpture, probably because of its less uniform character, though many of the remains of ancient sculpture that have been found in Attica are of Pentelic marble. It is curious that Hill remarks in reference to this marble, "The Pentelican . . . is now wholly unknown, and has been so for many ages."[26] The earlier statement of De Laet, "I don't remember that I have read about Pentelic marble anywhere else,"[27] also reflects the general state of ignorance in Western Europe concerning Greece under Turkish occupation.

Less is known with certainty about the nature of the marbles or other stones obtained in ancient times from the island of Chios. Pliny[28] states that in his opinion variegated marbles were first discovered in the quarries of Chios. From another passage,[29] where the reading is uncertain, some have inferred that Chian marble was uniformly black, though Mayhoff adopts another reading (*Melo*) in his edition of Pliny and does not think this refers to Chios. Theophrastus implies in the next section that the

[26] J. Hill, *Theophrastus's History of Stones* (London, 1746), p. 21.
[27] *de Pentelico non memini me alibi legisse*; see J. De Laet, *Theophrasti De lapidibus Graece et Latine cum brevibus annotationibus* (Leyden, 1647), p. 4.
[28] XXXVI, 46.   [29] XXXVI, 49.

THEOPHRASTUS ON STONES

Chian stone was black; he also says it was translucent, perhaps a reference to its surface appearance when polished. In section 33, where Theophrastus gives brief descriptions of inferior variegated dark stones used for seals, he apparently compares one of them with a stone from Chios. On the whole it would appear that the ancient Chian quarries yielded a black or dark marble or other rock, in which there were spots or streaks of light-colored minerals.

That the Theban quarries here mentioned by Theophrastus were located in the vicinity of ancient Thebes in Egypt rather than the Greek city of Thebes in Boeotia is almost certain, for not only does he go on immediately to name particular stones found near Thebes in Egypt, but he also says—what is well known—that the ancient Egyptian locality was celebrated for its great stone quarries. If Theophrastus is referring to quarries in or near the Egyptian Thebes, then he must be speaking of the limestone still so plentiful in this part of the Nile Valley. Even now there is evidence of the ancient workings there. If, as is more probable, Theophrastus meant to include the quarries in the general territory of Thebes, then sandstone and even granite could also be mentioned, although the great source of granite was at Syene (modern Aswan), about a hundred miles or so south of the city of Thebes. It is perhaps significant that the Theban stone which Pliny[80] mentions was apparently a granite.

6. *alabastrites.*

This was well known in ancient times, as is attested by numerous references to it in the works of early writers. Judging from the description given by Pliny,[81] this stone was in all probability compact stalagmitic calcium carbonate, the onyx marble of Egypt, sometimes called "oriental alabaster" to distinguish it from true alabaster, which is similar in appearance but is actually a compact variety of gypsum, a hydrated calcium sulfate. Even today this particular onyx marble is often loosely called "alabaster." It is a beautiful white or yellowish-white stone, slightly translucent, and frequently, though not always, traversed by bands of slightly differing shades. That this stone was extensively quarried in ancient

[80] XXXVI, 63, 157.   [81] XXXVI, 59-61.

COMMENTARY

Egypt is amply shown by the traces of the former workings as well as by the numerous objects composed of it that have survived, such as vases, statues, and even parts of buildings. However, here again it must be understood that Theophrastus is speaking in a vague general way of the territory of Thebes, for the nearest source of onyx marble appears to be considerably more than a hundred miles north of the site of the ancient city of Thebes. The ancient workings are still to be seen today, and the principal ones extend from Minia to Asiut.[82]

6. *chernites*.

This passage contains the only information we have about the nature of this stone. Pliny[83] mentions it but merely paraphrases the statement of Theophrastus. Hill[84] identified it as a white marble, though he did not explain his reasons. Against this identification is the fact that native marble was scarce and relatively little used in ancient Egypt. Furthermore, the few objects made of this stone that have been found are generally very small, by no means approaching a sarcophagus in size. Moore[85] identified *chernites* as true alabaster. The context suggests that it was either alabaster or a particular variety of Egyptian onyx marble; for Theophrastus commonly groups together mineral substances similar in nature or appearance. But it is not very likely that it was a true alabaster; such compact gypsum was scarce and little used in ancient Egypt, where only a few small objects made of it have been found. It appears more probable that it was a particular kind of onyx marble, possibly a white, more uniform variety, as contrasted with the common variety that was yellowish and banded. Theophrastus mentions particularly that a sarcophagus was made of *chernites*, and several such Egyptian burial objects composed of onyx marble have survived to our day.[86]

7. *poros*.

The usual spelling is πῶρος, not πόρος. Other Greek authors mention *poros* or *poros stone*, and their statements, together with

[82] A. Lucas, *Ancient Egyptian Materials and Industries* (London, 1948), pp. 75-77.
[83] XXXVI, 132. [84] *Theophrastus's History of Stones*, p. 23.
[85] N. F. Moore, *Ancient Mineralogy* (New York, 1859), p. 172.
[86] Lucas, *Ancient Egyptian Materials and Industries*, p. 463.

the evidence of geological formations and our knowledge of the stones used in ancient structures, show that these terms were general ones that were applied to rocks of cellular structure and low density such as calcareous tufa or fossiliferous limestone. The statement of Theophrastus here is somewhat reminiscent of the one made by Herodotus, who remarks,[37] in speaking of the building of the temple of Apollo at Delphi, that the front was of Parian marble and the main part of *poros stone*, which in this case was the calcareous tufa plentiful around Mt. Parnassos. Pausanias[38] states that the temple of Zeus at Olympia was made of *poros* locally obtained, undoubtedly the coarse fossiliferous limestone of that particular locality. In general, such limestone rock, being plentiful in Greece and easy to quarry or work, seems to have been a favorite building material, as is shown by its presence in a number of surviving ancient structures or ruins. The kind of *poros* which Theophrastus compares with Parian marble might be identified as travertine, a stone which in general appearance as well as chemical composition is very much like marble. There is ample evidence that travertine was widely used for building purposes in ancient times, particularly in Italy, and surviving examples show that fine-grained calcareous tufa or travertine was also frequently employed for statuary. But when the name *poros* is applied to a marble-like stone used by the Egyptians for elaborate buildings, it probably has a special meaning. It seems to mean some variety of Egyptian onyx marble, as this is the only marble-like stone that the Egyptians used in the construction of buildings; it was especially used in the construction of sanctuaries and temples, as is shown by surviving examples that have come down to us.[39] Though Theophrastus apparently speaks of onyx marble as *alabastrites* or *chernites* in the preceding section, it does not follow that this identification is wrong, since the stone occurs in several varieties that differ much in appearance. Moreover, the ancients often gave different names to the same mineral substance, or used the same name to denote two or more mineral species that we consider entirely distinct. The latter practice seems to have been true of *poros*, as it is very probable that this term

[37] V, 62. The text reads πωρίνου λίθου. [38] V, 10, 3.
[39] Lucas, *Ancient Egyptian Materials and Industries*, p. 75.

COMMENTARY

included not only calcareous tufa and fossiliferous limestone but also certain other soft rocks suitable for building purposes. The loose usage of the term *poros* by ancient authors, and the equally broad interpretation of its meaning by modern archaeologists, has been pointed out by Frazer.[40] Eichholz[41] rightly thinks that this *poros* must have been a special kind found in Egypt which had the lightness of the ordinary Greek *poros* but was not the same. This explains the difficulty in the text, which appears to say that *poros* has the lightness of *poros*.

7. *And a dark stone is also found in the same place, which is translucent like the Chian stone.*

Unless more than a word or two has dropped out between this and the preceding phrase, the reference is apparently to a dark or black stone found in Egypt which, at first sight, one might be inclined to identify as obsidian, since this is the only dark or black stone that exhibits any marked degree of translucency. Hill[42] believed that the stone was obsidian. But this identification is unlikely, since obsidian is not native to Egypt, although small amounts evidently were imported in ancient times to make ornamental objects, such as amulets and vases.[43] However, several kinds of dark-grey or nearly black stones were quarried there, as is shown by existing remains; the dark granite found at Aswan, for example, was used to some extent in the construction of buildings.[44] Since Theophrastus is dealing in this section with stones found in large masses, this might well have been the stone described here, though granite is certainly not translucent, unless the allusion is to the surface appearance of the polished stone. Basalt and diorite were also quarried in ancient Egypt. In addition, the Egyptians used a particular kind of diorite-gneiss, a banded or mottled black and yellowish-white rock, and also a black and white porphyry composed of white crystals imbedded in a black matrix.[45] On the whole, it seems likely that Theophrastus is allud-

[40] J. G. Frazer, *Pausanias's Description of Greece* (London, 1913), Vol. III, pp. 502-503.
[41] D. E. Eichholz, *Classical Review*, LVIII (1944), 18.
[42] Hill, *Theophrastus's History of Stones*, p. 24.
[43] Lucas, *Ancient Egyptian Materials and Industries*, pp. 473-74.
[44] *Ibid.*, p. 73.
[45] *Ibid.*, pp. 466-67, 474-75.

ing to one of these last two stones, especially since he compares the variety that he mentions with the Chian marble or stone, which, as indicated by the note on the preceding section, was probably a black rock variegated with spots or streaks of light-colored material.

8. *almost all those that can reasonably be cut and used as seals.*
The reading of the manuscripts, σχεδὸν ... λόγον, makes no sense. The meaning seems to be "almost all of those which," and this could be represented by σχεδὸν πάντες (or πλεῖστοι) τῶν. Here τῶν is added to go with γλυπτῶν. But this does not account for the presence of λόγον. De Laet and Hill changed this to λόγῳ τῶν, but the meaning of this is not clear. Schneider thought that the verb καταλέγονται might have survived as λόγον, and proposed σχεδὸν ὅσοι καταλέγονται εἰς τὰ σφραγίδια τῶν γλυπτῶν. The phrase κατὰ λόγον ("according to reason," "reasonably") is attractive, and Stephanides rightly included it in his emendation σχεδὸν πάντες οἱ κατὰ λόγον εἰς τὰ σφραγίδια γλυπτοί. Though it is impossible to know what was originally in the manuscripts, the emendation ⟨πάντες τῶν κατὰ⟩ seems to be a good one and has therefore been added to the text. These three words are of the right length to fill up the gap in the Aldine edition, but this does not confirm their accuracy, for the gaps are of different sizes in the manuscripts. In A there is room for only a few letters; the space in both B and C is longer, but not long enough. So nothing can be proved by the size of the gaps.

8. *And some are discovered in other stones when these are cut up.*
This apparently refers to crystals lining rock cavities or geodes, and especially to crystalline quartz which so commonly occurs in this manner, since in section 30 it is specifically stated that rock crystal and amethyst are found by dividing other stones.

9. *fire-resisting stones and millstones.*
Aristotle[46] refers to these two kinds of stones in a similar way except that he implies that they were fusible by themselves, whereas

[46] *Meteorologica*, IV, 6, 383B.

COMMENTARY

Theophrastus states here that they were fusible when placed in contact with certain other material. Since the two kinds of stones are mentioned together, it is likely that they were either similar in nature or had some common property or use. Though there is no direct evidence that would enable us to identify the first of these two classes of stones, several references in ancient literature make it possible to identify at least some of the class of millstones. Strabo,[47] for example, observes that the lava of Etna becomes millstone on cooling, and in another place[48] he mentions that an abundance of millstone was found on the volcanic island of Nisyros. Pliny,[49] on the other hand, states that a superior kind of quicklime was prepared from the stone commonly used for millstones (*molares*). Although it is very likely that a variety of stones were used for millstones in ancient times, the remarks of these authors indicate that they were usually made either from highly siliceous volcanic rocks or from common hard compact limestone. From the name it may be inferred that fire-resisting stones were varieties of limestone, and since Aristotle[50] mentions the fusibility of such stones immediately after referring to the manufacture of iron, this identification is even more likely; for limestone is generally added as an essential fluxing agent in the smelting of iron ore. Although there is no necessary connection between the sentence in Theophrastus dealing with these two stones and the two preceding sentences, it seems quite likely that such a connection was intended. If the passage is understood in this manner, then it also seems justifiable to assume that, because of the chemical nature of the two kinds of stones, the material that was burnt with them was the mixture of ore and fuel used in making up the charge for the smelting furnaces. For these two kinds of stones supply the two different types of fluxes needed for the reduction of ores, the acid type in the form of highly siliceous rocks and the usual basic type in the form of limestone. Though they lack precision, the brief allusions to ore-smelting in this section of Theophrastus are of considerable interest, since they are the earliest historical reference to the process.

[47] VI, 2, 3.   [48] X, 5, 16.   [49] XXXVI, 174.
[50] *Meteorologica*, IV, 6, 383B.

9. *become fluid along with the material.*
Some preposition like σύν ("with") is missing in the Greek text. Either ῥέουσι σὺν οἷς or συρρέουσιν οἷς might be expected.

9. *And some go so far as to say that all of them melt except marble and that this burns up and lime is formed from it.*
Though κονία has been translated as "lime," an alternate possibility is "ashes." This meaning is discussed in the notes on section 69. In that section Theophrastus describes briefly the "burning" of marble in order to make lime. He is essentially correct in noting the infusibility of marble, for most rocks, being composed of silica, silicates, or various mixtures of the two, melt at moderately elevated temperatures; but marble, which is a rock composed of nearly pure calcium carbonate, decomposes without melting at temperatures near 900° C. under ordinary atmospheric pressure. The residue, which is calcium oxide, is itself a very refractory substance that melts at about 2570° C. when it is pure; but this temperature was not available to the ancients and is not reached in modern furnaces operating on ordinary fuels. Modern investigators[51] have found, however, that calcium carbonate, either in the form of the pure compound or in the form of marble, does melt when both the temperature and the pressure are high enough.

10. *there are many which break and fly into pieces.*
Theophrastus means here that some stones are infusible because, like earthenware, they contain little or no moisture. According to Plato[52] the brittleness of earthenware is due to its mode of formation; for earthenware, like stone, is formed by the expulsion of water from a mixture of water and earth, followed by compression of the mass by the reaction of the surrounding air. However, in the formation of earthenware the mixture was thought to be so rapidly deprived of its water by the action of fire that the sudden violence of the compression made the product harder

[51] J. W. Mellor, *A Comprehensive Treatise on Inorganic and Theoretical Chemistry* (London, 1923), Vol. III, pp. 656, 836.
[52] *Timaeus*, 60D.

COMMENTARY

and more brittle than stone. Plato[53] also attempted to explain why an admixture of some water with earth was necessary in order to produce a material fusible by fire. He considered that if a body were composed of earth alone, not compacted by any unusual force, the interstices in it would be larger than the fire particles, which could therefore pass in freely without exerting any force that would tend to break up the mass. On the other hand, he considered that, if some water were present, the fire particles could force their way into the smaller pores of the water particles and thus break up or disturb these; and they in turn then acted on the earth, so that the entire mass was broken up and fusion was finally effected. But Plato believed that when the mass of earth was forcibly compacted, as was supposed to happen to earthenware, the pores were smaller and only fire particles could find an entrance. Theophrastus seems to imply here, as a logical extension of this argument, that the reason pottery and certain stones fly apart on the application of heat is that the fire particles force their way into the small pores of such bodies, which are thereby fractured owing to their inherent brittleness.

The negative οὐδ' should be bracketed in the text; the meaning ought to be that many stones, like pottery, fly into pieces owing to the action of fire. It seems most unlikely that Theophrastus intended to make an exception of pottery.

11. *they are useless unless they are ... wetted again.*
It is not obvious why the denser stones which harden on drying are supposed to become useless. Possibly the allusion is to their availability for cutting or carving; for some absorbent stones are definitely easier to work when wet. With sandstone, for example, this is certainly true; for it is cut much more easily when it is impregnated with water, as it often is in the quarries, than when it is completely dry. The statement that certain stones become softer and more brittle when they are dry is also somewhat obscure. There is a strong probability that the allusion is to native asphalt and related substances, which were well known and extensively used in ancient times.[54] These soften in a characteristic way when

[53] *Timaeus*, 61A-B.
[54] R. J. Forbes, *Bitumen and Petroleum in Antiquity* (Leyden, 1936).

subjected to moderate heat; and this would happen if they were exposed to the sun. This interpretation seems likely, since Theophrastus goes on immediately to discuss various brittle "stones" which were obviously natural bituminous substances.

12. *Some of those that can be broken are like hot coals when they burn, and remain like this for some time, such as those found in the mine at Binai which are brought down by the river.*

Bina was a town in Thrace, but its exact location is unknown. Theophrastus uses the plural form *Binai*. The account given by the writers of the *Etymologicum Magnum*, namely, that the place received its name because of the immorality of its inhabitants (βινεῖν), is obviously false. Procopius[55] mentions a castle βίνεος in a list of forts, and this may have been at the place in question. Stephanus of Byzantium[56] lists *Benna* as a city of Thrace and states that the spelling was sometimes *Bena* or *Beina*, though the former was better. At the time when Stephanus lived, *Beina* and *Bina* would have had the same pronunciation, so that it seems likely that this was the place which Theophrastus mentions here. Unfortunately, Stephanus, too, fails to locate the city. That it was probably on a Thracian river named Pontus is indicated by a passage in the pseudo-Aristotelian work *De Mirabilibus Auscultationibus*.[57]

The allusion here is undoubtedly to some sort of natural solid bituminous substance, though the description is not adequate for an exact identification. The fact that the stones are described as brittle would suggest a bitumen associated with shale or soft limestone. The most likely identification is rock asphalt; for the combustible stone of Erineas mentioned in section 15 was probably rock asphalt too, and this is said to be like the kind found at Bina. The material called *spinos*, mentioned in the next section, was probably a solid bitumen, and this also suggests that the stones found at Bina were asphaltic in nature. The only objection to such an identification is that bitumens are not of frequent occurrence within the boundaries of ancient Thrace. There is a possibility

[55] *De Aedificiis*, IV, 4.   [56] *S.v.* Βέννα.   [57] Sec. 115.

COMMENTARY

that the "stones" found at Bina were lignite, or some related non-asphaltic pyrobitumen, though the fact that Theophrastus mentions lignite separately in section 16 makes this identification somewhat unlikely.

13. *spinos*.

The only other ancient work in which this mineral substance is named seems to be the *De Mirabilibus Auscultationibus*. Here it is briefly mentioned in two separate passages. The passage in section 33 may be translated as follows: "And in Thracian Bithynia the so-called *spinos* is found in mines, and they say that fire is kindled from it." And section 41 states that *spinos* burns when it is cut up and put together again and sprinkled with water.

Since Bithynia was the name of a province in Asia Minor, it might seem that a geographical contradiction exists in the first of these two passages, but it is clear from an account of Strabo[58] that there were Bithynians living in Thrace, and in fact the inhabitants of the province of Bithynia originally came from Thrace to Asia Minor. Both these passages show, therefore, that *spinos* was a combustible mineral substance found in Thrace.

The behavior of *spinos* when moistened with water suggests that it was the same mineral substance as the Thracian stone mentioned by various ancient authors, such as Nicander,[59] Dioscorides,[60] and Pliny.[61] Hence *spinos* appears to have been an early name for Thracian stone. Though it is impossible to identify it with any degree of exactness, *spinos* or Thracian stone was probably some kind of asphaltic bitumen. Some have suggested that Thracian stone was lignite or brown coal,[62] and others have identified it as ordinary bituminous coal,[63] or, in an attempt to account for its peculiar behavior with water, as coal containing pyrite.[64] The conjecture of Stephanides[65] that it was an asphaltic

[58] XII, 3, 3.
[59] *Theriaca*, 45.
[60] V, 146 (Wellmann ed., V, 129). Wherever the *Materia Medica* of Dioscorides is mentioned, the first reference is to the German translation of Berendes and the second is to Wellmann's edition.
[61] XXXIII, 94.
[62] J. D. Dana, *System of Mineralogy* (New York, 1909), p. 1024.
[63] Forbes, *Bitumen and Petroleum in Antiquity*, Table I.
[64] H. O. Lenz, *Mineralogie der alten Griechen und Römer* (Gotha, 1861), p. 18.
[65] M. K. Stephanides, *The Mineralogy of Theophrastus* (in Greek), (Athens, 1896), p. 211.

lignite is probably nearer the truth. It is not improbable that *spinos* and the combustible "stones" found at Bina were merely varieties of the same mineral substance, different perhaps in respect to asphalt content, or possibly only in superficial appearance. That the ancients confused the various sorts of solid natural bitumens is fairly certain; even in modern times their classification has been a difficult problem.

The origin of the word *spinos* is obscure, and even the quantity of the first vowel is uncertain. It seems better to write σπίνος in the text, following Bekker in his edition of Aristotle, rather than σπῖνος, which Wimmer prefers.

13. *in mines.*
The actual reading of the manuscripts is ἐν τοῖς μετάλλοις ("in the mines"), but Wimmer changed this to τοῖς (αὐτοῖς) μετάλλοις ("in the same mines"), meaning the mine at Bina mentioned in section 12. This emendation is supported by the passage in *De Mirabilibus Auscultationibus* which states that *spinos* was found in mines in a certain part of Thrace, by the fact that Bina was located in Thrace, and by the apparent similarity of *spinos* and the "stones" found at Bina. Nevertheless, it by no means follows that *spinos* was mined at Bina, and Wimmer's emendation has not been adopted.

13. *If this is cut up and the pieces are piled in a heap, it burns when exposed to the sun, and it does this all the more if it is moistened and sprinkled with water.*
This statement probably describes the spontaneous combustion of a pile of bituminous material. Dry piles of such materials often ignite spontaneously under the proper conditions, as has been observed repeatedly with unventilated piles of ordinary coal. The last part of this statement, which is probably based on uncritical observation, describes the effect of throwing water on a pile of smoldering bituminous material. The clouds of smoke and steam are regarded as a sign of increased combustion. All the other ancient writers who describe the Thracian stone, some of whom certainly made use of this passage in Theophrastus, entirely misunderstood the reasons for its combustion, as is evident from their

COMMENTARY

accounts. Nicander[66] states that Thracian stone flames up when it is moistened with water and is quenched by oil. Dioscorides[67] increases the marvelous nature of the phenomenon by stating that the stone was ignited by water and quenched by oil. Pliny[68] discusses the heat that is developed when water is added to Thracian stone, which he compares with lime, and also says that it could be quenched with oil. Evidently the story increased with the telling. Theophrastus was apparently the first to mention spontaneous combustion. Moreover, he seems to have been the only ancient writer to allude to this phenomenon in a reasonably clear and rational manner.

14. Liparean stone.

This was evidently named from the volcanic group, called at present the Lipari Islands, lying off the northeastern coast of Sicily. These islands are still the scene of much volcanic activity. The locality and the description leave little doubt that this so-called Liparean stone was what we now call obsidian. Large quantities of this dark volcanic glass occur at certain places on these islands. The mention of pumice in connection with the Liparean stone supports the identification, for both these varieties of glassy rhyolite or liparite commonly occur together, often in the very way described by Theophrastus.

Stephanides,[69] on the basis of a very literal interpretation of the statements in this section, identified this Liparean stone as a combustible mineral substance, possibly a volcanic rock impregnated with asphalt, but it is clear from what Theophrastus says in sections 19 and 20 about the creation of pumice by fire that combustion in the modern sense of the word is not to be understood here. The ancients apparently made little or no distinction between actual burning and the phenomena connected with molten material at a high temperature. Thus when pumice was formed by the expulsion of gases from molten volcanic glass at the time of solidification, this did not seem to differ from the combustion of a mineral substance such as lignite, especially since the

[66] *Theriaca*, 45. [67] V, 146 (Wellmann ed., V, 129).
[68] XXXIII, 94.
[69] *The Mineralogy of Theophrastus*, pp. 211-12.

end products, pumice in the one case and ash in the other, are somewhat similar in superficial appearance.

Obsidian from the Lipari Islands and the neighboring volcanic regions was used from early times for ornamental and useful purposes by the peoples of the Mediterranean region, as is clearly shown by the numerous archaeological finds.[70]

Theophrastus appears to be the first writer to give a distinctive name to obsidian, though it is not improbable that the black stone mentioned by Plato[71] was also obsidian. Pliny[72] called it *obsiana* (neuter plural); this spelling appears in Mayhoff's text, but there is a variant reading, *obsidiana*, which is the origin of the present English name.

14. Melos.

This is an island of volcanic origin in the southern Aegean about halfway between Crete and the southern tip of Attica. In his reference to the pumice of Melos, Theophrastus seems to mean that it occurred in separated cells in the solid rock, though this was not obsidian. Probably he refers to the occurrence of pumice in ordinary rhyolite.

Pumice is abundant both on the Lipari Islands and on Melos, and these localities are leading commercial sources at the present day. Theophrastus refers to pumice in more detail in sections 19, 20, 21, and 22. See also the notes to these sections.

15. Tetras.

It is clear from the reference to Lipara that this was situated somewhere in the northeastern corner of Sicily. This locality is not mentioned elsewhere. Though it is very brief, the statement about the stone found at Tetras shows clearly that it was some volcanic product similar to the one mentioned in the preceding section.

15. Erineas.

This name is not found elsewhere. Strabo[73] mentions a town,

---

[70] H. Blümner, *Technologie und Terminologie der Gewerbe und Künste bei Griechen und Römern* (Leipzig, 1875-1887), Vol. III, pp. 273-74; J. R. Partington, *Origins and Development of Applied Chemistry* (London, 1935), pp. 103, 324.
[71] *Timaeus*, 60D (τὸ μέλαν χρῶμα ἔχον εἶδος).
[72] XXXVI, 196.   [73] IX, 4, 10.

COMMENTARY

Erineos, which was in Doris in central Greece, but evidence for connecting it with the promontory called Erineas is lacking. Though the text does not actually say so, it seems probable that, like Tetras, this was a place in Sicily. The similarity of the stone found at Erineas to those at Bina, the odor on burning, and the appearance of the residue after combustion, all tend to show that it was a bituminous product such as rock asphalt. The occurrence of great quantities of this material in Sicily lends considerable support to this identification. The deposit at Ragusa, for example, which forms a bed 10 to 50 feet thick and 1,600 to 2,000 feet long, is one of the largest in Europe, and, in spite of the fact that it has been worked for a long time, recently over 100,000 tons of rock asphalt have been obtained from it annually. Smaller deposits of commercial importance occur at Modica and Scicli in the same region. The rock asphalt in these localities is a soft fossiliferous limestone containing from 2 to 30 per cent of actual asphalt.[74] If one assumes that Erineas was in Sicily, the combustible stone found there could not have been some non-asphaltic pyrobitumen such as coal or lignite, for these do not occur on the island.[75]

15. the stone found at Binai.
The reading τῷ ἐν Βίναις is the emendation of Turnebus. The manuscripts have ταῖς κίναις, but this makes no sense. Binai has already been mentioned in section 12.

16. Among the substances that are dug up because they are useful, those known simply as coals are made of earth, and they are set on fire and burnt like charcoal.
Here Theophrastus mentions Liguria, a coastal district in northwestern Italy, and Elis, the district in Greece in the northwestern part of the Peloponnesus, where Olympia is situated. Some commentators on this interesting passage have concluded that Theophrastus is referring to anthracite or to bituminous coal. Certain considerations, however, make it very improbable that this conclusion is justified. One is that true coal does not occur in Greece,

---

[74] H. Abraham, *Asphalts and Allied Substances* (New York, 1945), Vol. I, pp. 229-34.
[75] W. McInnes, D.B. Dowling, and W.W. Leach, *The Coal Resources of the World* (Toronto, 1913), Vol. II, pp. 721-33.

and, though small deposits of anthracite are to be found in the Western Alps in the extreme north of Italy, the most abundant type of non-asphaltic pyrobitumen that is found in the region once known as Liguria is lignite, and this is also of fairly common occurrence in Greece.[76] Some important evidence is also given by Theophrastus himself in his treatise *On Fire* (sec. 75), where he explains why prepared coals (i.e., charcoal) are blacker than the kind that are dug up. This seems to indicate clearly that the latter was brown coal or lignite. The allusion in the present passage to the earthy nature of these coals leads to the same conclusion. It is interesting to note that at the present time lignite is mined on a large scale for domestic and industrial purposes in the same part of Italy. Large quantities are also mined in Greece.

Even though this passage does not refer to the use of true coal, it is nevertheless of historical importance as containing the earliest known account of the use of a mineral product for fuel. The use of lignite is scarcely mentioned elsewhere in ancient literature. Apart from Antigonus of Carystus,[77] who says that according to Theopompus coals were dug up for use in the neighborhood of the Thesprotians, Theophrastus appears to be the only ancient writer who touches on the subject. Hence it seems almost certain that lignite was not commonly employed in ancient times. From what Theophrastus says, it is not certain to what extent lignite was used as a general fuel, since he merely states that metal workers made use of it. They probably found it of special value for the operation of forges and furnaces. Modern writers on ancient technical processes have often assumed that wood and wood-charcoal were the only fuels available and in use among the Greeks and other ancient peoples, but this passage gives unquestionable evidence of the use of mineral fuel, at least in certain industrial arts.

16. amber.

This is mentioned again in section 29, where it is said to be found in Liguria, but that is certainly incorrect. See the notes on section 29 for a discussion of this question.

[76] *Ibid.*
[77] CLXX (186).

COMMENTARY

17. *In the mines at Scaptē Hylē a stone was once found which was like rotten wood in appearance. Whenever oil was poured on it, it burnt, but when the oil had been used up, the stone stopped burning, as if it were itself unaffected.*

Scaptē Hylē was a mining district in Thrace opposite the island of Thasos in the Northern Aegean. According to Davies[78] the modern Eski Kavala is perhaps the district. In Wimmer's text the name appears in the genitive case and is written as one word (Σκαπτησύλης), but the true nominative form is Σκαπτὴ Ὕλη ("a forest that may be dug"). In the manuscripts the name appears as two words, ἐγκαπτῆς ὕλης, and the first includes the preposition ἐν. Turnebus changed this to ἐν σκαπτησύλης. The Latin name is Scaptesula, but Scaptensula is the spelling found in Lucretius.[79]

Something seems to be lacking in this passage. Does Theophrastus mean that the stone became ignited as soon as oil was poured upon it, or does he mean that when oil was poured upon the stone and ignited, it then burnt away, leaving the stone in its original state? The second meaning certainly seems more probable, though the first one may well have been what Theophrastus intended; for when ancient authors say that Thracian stone and other combustible mineral substances are ignited by water and extinguished by oil, they seem to regard this as a phenomenon worthy of special mention, because it is opposed to the normal order of things. It is untrue, however, that any mineral can be ignited by the mere act of pouring oil upon it. If such a notion was held by Theophrastus and other ancient writers, it probably originated from distorted hearsay evidence or from false reasoning divorced from experience.

Moore[80] thought that Theophrastus was really referring to asbestos. The color of the stone makes this unlikely, though its structure makes it less improbable, since some forms of decayed wood do have a fibrous structure like asbestos. We know from statements of various early authors that asbestos was known in an-

[78] O. Davies, *Roman Mines in Europe* (Oxford, 1935), p. 235.
[79] K. Lachmann, *In T. Lucretii Cari De Rerum Natura Libros Commentarius* (Berlin, 1882), p. 395 (on Lucretius 6, 810).
[80] Moore, *Ancient Mineralogy*, p. 153.

tiquity, and that it was mainly used for the manufacture of incombustible cloth, though evidently wicks for oil lamps were also made of it.[81] Moreover, direct evidence of the use of asbestos by the ancients has been obtained in modern times by the discovery of ancient garments woven from this mineral.[82] It is, however, unlikely that Theophrastus is alluding to asbestos, since the mineral does not occur in the locality mentioned. There were only two known sources of asbestos in Greece and its vicinity in ancient times: Karystos at the southern extremity of the island of Euboea, and a place to the southeast of Mt. Troodos on Cyprus, where the abandoned workings are still to be seen today.

It is much more probable that Theophrastus is referring to the well-known brown fibrous lignite, which in appearance and in other respects very often closely resembles rotten wood. Lignite of various kinds is known to occur in the region named by Theophrastus. He seems to be pointing out that when oil is poured on this material and ignited, the oil burns away without igniting the material, though this would be combustible under the proper conditions. Lignite of the kind to which he apparently refers often contains in its natural state as much as 20 per cent of water; thus it cannot readily be ignited, though it is combustible when it is properly dried out, and this soon happens if, for example, it is placed on a bed of glowing coals. As Theophrastus shows in the last sentence of this section, he is dealing here and in most of the preceding sections of this chapter with mineral substances that are actually combustible. The discussion of incombustible minerals is taken up in the next chapter and, if the interpretation of this passage is correct, his description of a combustible mineral substance which under certain conditions is incombustible affords a logical transition to his next general topic. Indeed, since Theophrastus in other places in this treatise makes similar transitions, this peculiarity of his style might possibly be taken as additional evidence in support of this identification.

17. *Whenever oil was poured on it, it burnt.*
In this passage the optatives ἐπιχέοιτο and ἐκκανθείη, which are

---

[81] Strabo, X, 1, 6; Dioscorides V, 155 (Wellmann ed., V, 138); Pliny, XIX, 19-20, and XXXVI, 139; Plutarch, *De Defectu Oraculorum*, 434A; Pausanias, I, 26, 7.
[82] Stephanides, *The Mineralogy of Theophrastus*, p. 121.

COMMENTARY

used with ὅτε, are followed by verbs in the present tense instead of the imperfect, which would be usual. The manuscripts have ἐκκαίεται, which Aldus changed to καίεται, and τότε παύεται. It would be possible to emend the text and to read ἐκαίετο and τότ' ἐπαύετο, but it is not certain that Theophrastus would have felt obliged to follow the strict sequence of tenses. It is, however, necessary to translate ἐκκαίεται as "it burnt" rather than "it burns." It seems better to restore ἐκκαίεται, the reading of the manuscripts, especially since ἐκκαυθείη follows; here Wimmer has taken the reading of Aldus.

18. *anthrax*.

This appears to have meant originally a glowing live coal; the word was used later, as Theophrastus uses it here, to mean a transparent precious stone of a deep red color. It appears to have been first used as the name of a gem by Aristotle, who says that "the seal-stone called anthrax is the least ⟨affected by fire⟩ of all the stones."[83] Theophrastus, however, is the first to give descriptive details by which the stone can be identified. Though *anthrax* was probably a generic term that could have been applied equally well to the ruby, red spinel, or red garnet, it is fairly certain from the evidence available that the stone designated by this name was nearly always red garnet at the time of Theophrastus. In the first place, no engraved rubies or spinels dating from the Hellenistic period are known with certainty, whereas many engraved garnets have come down to us and exist today in various museums.[84] In the second place, the ruby, with its high degree of hardness, could not normally have been used for seals by the Greeks, since they would have experienced great difficulty in engraving this stone with the abrasives then available. Garnet, on the other hand, with its lower degree of hardness, offered no such technical difficulty.

It is worth noting that when Theophrastus begins to discuss incombustible stones, he mentions a variety which seems to be related to combustible stones by its name and appearance, and thus he makes an easy and not illogical transition from one class

[83] *Meteorologica*, IV, 9, 387B (17).
[84] Blümner, *Technologie und Terminologie der Gewerbe und Künste bei Griechen und Römern*, Vol. III, p. 245; A. Furtwängler, *Die antiken Gemmen* (Leipzig and Berlin, 1900), Vol. II, pp. 130-46, 153-73.

to the other. It is clear from the brief remark of Aristotle quoted in the preceding paragraph that Theophrastus was not the first to notice that *anthrax* could not be burnt. It is easy to understand why philosophers of the Peripatetic school would emphasize the paradox involved in a stone, which, though connected with fire by its name and appearance, was itself incombustible. As a matter of fact, garnet, though quite incombustible, is more readily changed by fire, owing to its lower melting point, than the different varieties of quartz from which most of the seals were made at the time of Theophrastus.

18. *One might say that it has great value; for a very small one costs forty pieces of gold.*

Probably the reference is to gold staters of Alexander III or his father Philip II of Macedon, which were in common use at the time; each of them weighed about 8.6 grams but had a much greater purchasing power than modern coins of the same precious-metal content. King[85] thought that in an age of extended commerce such a high price could scarcely have been paid for a stone as common as garnet, and it was largely because of this that he identified this first *anthrax* mentioned by Theophrastus as the ruby. However, engraved garnets first appeared in the Hellenistic period, as is shown by the surviving examples; possibly they were introduced during the lifetime of Theophrastus, and since they may have been a scarce and highly prized novelty at the time of their introduction, the price mentioned does not seem excessive. Furthermore, flawless garnets of brilliant red color are much scarcer than good specimens of the various kinds of quartz that the ancient Greeks valued highly as precious stones. It should not be forgotten that stones which were highly prized by the ancients, although apparently very costly at the time, would not usually be termed precious today; most of them would be rated as semiprecious, or even less valuable, stones.

18. *Carthage and Massalia.*

Since Carthage in North Africa and Massalia at the site of modern Marseilles were both important seaports, it must be under-

---

[85] C. W. King, *Natural History of Precious Stones and of the Precious Metals* (London, 1870), p. 225.

COMMENTARY

stood that these were merely points of export, not the localities where the garnets were found. Coraës[86] suggested that Μασσυλίας, a district in the country behind Carthage, should be the correct reading here. Though it is true that Strabo mentions only localities in North Africa as western sources of these precious stones, the more comprehensive list of Pliny[87] includes Massilia (Marseilles) and Olisipo (Lisbon) as well. Theophrastus mentions Massalia again in section 34 as a place from which precious stones were obtained. On the whole, the conjecture of Coraës is not plausible enough to justify an alteration of the text.

19. *The stone found near Miletus does not burn; it is angular and there are hexagonal shapes on it. It is also called anthrax, and this is remarkable, for in a way the nature of adamas is similar.*

The kind of *anthrax* found near Miletus on the western coast of Asia Minor evidently had a striking and peculiar form. Theophrastus appears to be speaking of a well-crystallized mineral with hexagonal facets. It is significant that both garnet and spinel often occur crystallized in this manner. Of the two, spinel seems more likely, since the *anthrax* mentioned in the present section is apparently different from the kind described in section 18, which was almost certainly garnet. Since the *anthrax* of this passage is said to be like *adamas*, it is even more likely that it was spinel. Though *adamas* seems to have been a general term used for several minerals that were unusually hard, the descriptions of Pliny[88] suggest that it generally referred to corundum, particularly the mixture known as emery; this frequently contains spinel in addition to corundum, and almost invariably magnetite, which is very similar to spinel in crystal form. The occurrence of large deposits of emery in Asia Minor not far from the site of ancient Miletus supports the relationship between this kind of *anthrax* and *adamas*. Most of the emery in these deposits contains only about 50 per cent of corundum, and the remainder consists of the associated minerals.[89] Though it is remarkable that this particu-

[86] Ed. of Strabo, IV, 357.   [87] XXXVII, 92-97.
[88] XXXVII, 55-61; see also the notes on sec. 44.
[89] C. Schmeiszer, *Zeitschrift für praktische Geologie*, XIV (1906), 188.

lar mineral should have been called *anthrax*, the explanation may be that spinel not only occurs in the red transparent form, to which this name was applicable without question, but also in dark or black varieties that resemble the magnetite which is often the most conspicuous component of emery. However, when *anthrax* is said to be similar to *adamas*, the reference may be, not to form or color, but rather to hardness, the special property for which *adamas* was noted. That would also identify this type of *anthrax* as spinel rather than garnet, since the hardness of spinel is close to that of emery, whereas the hardness of garnet is distinctly lower.

19. *This power of resisting fire does not seem to be due to the absence of moisture, as is true of pumice and ashes. For these cannot be set on fire and burnt, because the moisture has been removed.*

According to certain theories of Aristotle[90] which Theophrastus appears to follow closely here, stones like *anthrax* are incombustible because they contain no moisture and hence lack pores of the proper size to admit fire. For much the same reason pumice and ashes are also incombustible; the difference is that they are produced from materials which originally contained moisture, whereas incombustible stones are free of moisture from the beginning.

19. *pumice.*

In the manuscripts this word appears as κίσσηρις, except in one place in this section where it is κίτηρις. The correct form is κίσηρις, which is used by all authors except Theophrastus. The descriptions in sections 20, 21, and 22 indicate that the word sometimes denoted certain cellular or friable rock material that would not now be called pumice, but the localities that are mentioned show that ordinarily the term had the same meaning that the word pumice has today, and for that reason it has been so translated. Pliny[91] shows that the corresponding Latin word *pumex* also had a slightly broader meaning than the modern term, though it is

[90] *Meteorologica*, IV, 9, 387A.   [91] XXXVI, 154.

COMMENTARY

equally clear from his account that the word normally had the same significance.

19. *some think that pumice is formed entirely as a result of burning.*
The ancients apparently made little or no distinction between true combustion and other high-temperature phenomena, as was mentioned in the notes on section 14. Evidently fire was a term that included all phenomena involving light and a high temperature. Therefore, when Theophrastus speaks of the origin of pumice from burning, combustion is not to be understood, but rather the formation of this material in the usual way by the expulsion of gases from molten lava. Superficially, of course, this process sometimes closely resembles actual combustion, especially when, as often happens, the gases evolved in the volcanic action take fire.

19. *with the exception of the kind that is produced from the foam of the sea.*
The pumice that was thought to be produced from foam is clearly the same as the floating pumice still found around the shores of islands in the Aegean Sea. Such pumice emanates from the active volcanic island of Thera (Santorin), where considerable quantities are to be seen floating on the surface of the water. Theophrastus evidently believed that it was formed in some way from the foam of sea water. This notion may have been obtained from Aristotle, for in a treatise *On Plants* usually ascribed to the latter, a theory is advanced to explain how floating stones could be formed from sea water.[92] According to the author[93] of this treatise, such stones were produced by the violent collision of one wave with another. First of all, foam is produced which congeals with the consistency of oily milk. When the water is dashed against sand on a beach, the sand collects the fat part of the foam, which dries with an

[92] *De Plantis*, II, 823B.
[93] Though this treatise is included in the corpus of Aristotelian writings, there is considerable doubt that Aristotle was the actual author. It appears to be a later production of the Peripatetic school, and there is some possibility that Theophrastus, or a pupil of Theophrastus, was the real author.

excess of salt from the water, so that the particles of the sand cohere and ultimately become stones.

20. *the porous stone which changes to pumice when it is fired.*
Both the name and the nature of this stone are uncertain. Wimmer retains the manuscript reading διαβάρου, which is accepted by all the editors except Hill, who substituted the word Ἀραβικοῦ (printed incorrectly with a rough breathing) and translated it as "Arabic stone," which according to Dioscorides[94] and Pliny[95] is a substance resembling ivory. There seems to be little justification for this conjecture. Since Theophrastus is discussing pumice and similar cellular stones, it seems probable that there is merely a slight error in spelling and that the reading should be διαβόρου, a word meaning "porous." The stone itself may have been a volcanic tufa of some kind.

The negative οὐ, which occurs before κισσηροῦται in the Aldine edition but is not in the manuscripts, is a difficult reading to interpret. If it is accepted, the passage refers to a stone which "does not change to pumice when it is fired." However, the οὐ in the Aldine edition is in the form of an abbreviation of the same size as the relative pronoun ἥ ("which"); this occurs in manuscripts B and C, though ἤ is written with a smooth breathing in manuscript A. Most of the editors prefer ἣ καί (i.e., "which also changes"), but Wimmer accepts the reading of Aldus. Since this is the more difficult reading and therefore harder to explain, it is easy to see why Wimmer felt obliged to adopt it. Actually, the evidence of the manuscripts seems to carry more weight, and the acceptance of the reading ἥ would make it easier to understand the meaning of the passage. There are many misprints in the Aldine edition; it is quite possible that the abbreviation representing οὐ is not correct, and there is no evidence that Aldus had another manuscript from which he might have obtained this reading. It seems best to accept ἥ, since it appears in two of the manuscripts. If ἤ ("or"), which appears in A, is correct, then some other verb must have dropped out before it; but it is possible that the smooth breathing is a mistake, and certainly A provides no evidence of a

---

[94] V, 148 (Wellmann ed., V, 131).   [95] XXXVI, 153.

COMMENTARY

negative. The genitive τῆς φλογουμένης is also a problem, since ἣ φλογουμένη ("which when it is fired") might be expected.

20. *for pumice is found especially in places that* . . .
The phrase ἐν τοῖς is incomplete. The meaning seems to be "in places like this," i.e., in craters of volcanoes. Schneider suggested ἐν τοῖς (καιομένοις), which would mean "in places that are on fire." Turnebus gives *in ardentibus* as a Latin translation. A word ending in -τοις might easily have dropped out after ἐν τοῖς. Thus πυρικαύστοις would mean "places that have been subjected to burning." Other possibilities are ἐν τούτοις ("in these places") or ἐν τοῖς τοιούτοις ("in such places").

21. *Nisyros.*
This is an island of volcanic origin in the southern Aegean near the coast of Asia Minor. Eruptive rocks of various sorts are found there. From the description it seems likely that the material that was found was a loosely compacted volcanic tufa or ash rather than a pumice, though the reference may be to an especially soft and friable kind of pumice.

21. *The kind found in Melos is all . . . but some.*
Since an adjective seems to be missing after πᾶσα μέν, Schneider suggested that ἔνια δ᾽ αὖ might be changed to εὔθραυστος ("easily broken"), followed by ἐν λίθῳ δέ; but Stephanides proposed σχεδὸν ὡς ἐν Νισύρῳ ("almost as in Nisyros"), followed by ἔνια δ᾽ αὖ.

22. *They differ from one another in color, density, and weight.*
It should be noted that the various kinds of pumiceous or scoriaceous rocks are distinguished only by their physical characteristics, for no hint is given that Theophrastus may have considered possible differences in composition.

22. *the kind that comes from the lava stream in Sicily is black.*
This black variety probably would not be termed pumice in our modern system of classification, but rather volcanic scoria. The article ⟨ἡ⟩ is needed and has been added to the text.

22. *the malodes.*

The word μαλώδης, which is the reading of the manuscripts, does not occur elsewhere. It is possible that it is a mistake for μηλώδης, meaning "quince-yellow," and could be called "a pale yellow kind," though such an adjective should have the article ἡ before it. It was evidently taken so by Stephanides,[96] who translates it into Modern Greek as ὑποκίτρινος. Pumice made yellow by disseminated sulfur is a material of common occurrence in Sicily and the neighboring volcanic islands. However, there have been other interpretations. Turnebus changed the word to ἁλμώδης,[97] Furlanus preferred μυλώδης ("a millstone"), and Schneider suggested ἡ Μηλία ("the Melian stone"). Stephanides[98] finally decided in favor of ἡ μυλώδης, but it is hard to say which is correct. It seems best to add ⟨ἡ⟩ to the text.

22. *The one that comes from the lava stream can cut better than the white kind.*

This is another indication that the black "pumice" was volcanic scoria and the white kind our ordinary pumice. It is interesting to note that Theophrastus, when he refers to the practical value of pumice, alludes only to its use as an abrasive. Later ancient writers emphasize medicinal uses. This explains why white pumice is here considered a less desirable kind, whereas such writers as Dioscorides[99] and Pliny[100] state specifically that the best pumice is recognized by its white color, its lightness, and the ease with which it can be powdered.

22. *but the kind that comes from the sea itself cuts best of all.*

This is another reference to the floating pumice mentioned in section 19 and discussed in the notes on that section. The article ⟨ἡ⟩ has been added to the text.

[96] *The Mineralogy of Theophrastus*, p. 111.
[97] This word is used by Theophrastus for soil "impregnated with salt" (*History of Plants*, VIII, 7, 6).
[98] M. K. Stephanides, *Athena*, XIV (1902), 368.
[99] V, 124 (Wellmann ed., V, 108).
[100] XXXVI, 154-56.

COMMENTARY

22. *But we must consider elsewhere the causes of the differences between stones that either burn or do not burn.* This vague promise or cross reference is similar to the one at the end of section 38 and very similar to the one at the end of the last section of the treatise *On Fire.* No further treatment of combustible and incombustible stones is to be found in any other extant work by Theophrastus. He does, however, in his *History of Plants,*[101] give some additional information about pumice.

23. *sardion, iaspis, and sappheiros.* For the identification of these stones, see sections 30, 27, and 37 respectively.

23. *smaragdos.*
Though the word "emerald" is derived from the Greek σμάραγδος, which has often been translated in this way, the accounts of early writers show clearly that in ancient times various stones of pronounced green color were listed under this name. The statements of Theophrastus make it doubtful whether true emerald was even known to him, and there appears to be no certain evidence on other grounds of its use among the Greeks. On the other hand, the more detailed descriptions of Pliny[102] indicate that emerald was known as *smaragdus* in his day, and archaeological discoveries afford ample proof that the Romans made use of this precious stone.[103] But Pliny,[104] who states that there were twelve different kinds of *smaragdus* and describes some of these varieties, makes it clear that a number of minerals other than the particular variety of beryl called emerald were included under the ancient name. Though it is impossible to determine with certainty what all of these were, probably any transparent or translucent green mineral that resembled emerald, even one as common as green quartz, would have been classified under *smaragdus*, and Pliny's descriptions and the localities that he mentions indicate that certain copper minerals, such as malachite, were classified in this way. It is also probable that imitation green stones composed of glass or stained

[101] IX, 17, 3. [102] XXXVII, 62-73.
[103] Blümner, *Technologie und Terminologie der Gewerbe und Künste bei Griechen und Römern*, Vol. III, p. 239.
[104] XXXVII, 65.

· 97 ·

rock crystal were given the same generic name. Pliny[105] alludes to the existence of books which contained directions for staining quartz in imitation of *smaragdus* and other precious stones, a fraud, he remarks, that was more lucrative than any other. Seneca[106] also mentions the staining of stone to resemble *smaragdus*. The *Stockholm Papyrus*, which contains numerous recipes for the imitation of precious stones by the staining of rock crystal, shows that there was a substantial basis for these remarks. According to this papyrus,[107] green precious stones were counterfeited by applying copper salts and organic coloring materials to quartz after its surface had been roughened. Rock crystal colored in this way could not have passed for a clear transparent green stone like emerald, though counterfeit stones of this kind may well have been a tolerable imitation of translucent green quartz. Whether such imitations passed for natural stones is uncertain, but the wording of the recipes indicates, at least, that they were known by the name *smaragdos* without qualification. Since the recipes given in the papyrus for imitating this particular stone are about equal in number to all the recipes for imitating other kinds of stones, it is clear that these counterfeit green stones were frequently used by the ancients.

23. *it makes the color of water just like its own.*

This supposed property of *smaragdos* is not mentioned by any other ancient writer, though Pliny[108] in a somewhat analogous passage remarks that from a distance such stones appear larger than they really are, because their green color is reflected by the surrounding atmosphere. Some commentators have supposed that this statement of Pliny was based on a misinterpretation of the present passage in Theophrastus, and King[109] even supposed that Pliny's account represents the original sense of the Greek passage, and that ὕδατος ("water"), which now appears in the text, is a corrupt reading for ἀέρος ("air"). Actually, the statement of Pliny is so different that it is unlikely that he paraphrased or even used

[105] XXXVII, 197.   [106] *Epistulae Morales*, 90, 33.
[107] O. Lagercrantz, *Papyrus Graecus Holmiensis* (Uppsala, 1913), pp. 7-8, 11-13, 14, 19-21, 22, 23-24, 165, 174, 176, 177, 179, 182-83, 193, 194, 195, 196-97, 199-200.
[108] XXXVII, 63.
[109] *Natural History of Precious Stones and of the Precious Metals*, p. 280.

Theophrastus at this point. When Theophrastus says that water is colored by *smaragdos*, his statement apparently has a rational basis, for under proper conditions bright green stones do impart a greenish cast to the water in which they are submerged. The phenomenon is best seen when the illumination is oblique and the stone is placed in a small opaque white vessel. It is rather curious that the same property is not ascribed to precious stones of other colors, for these can also impart their color to water through reflection, especially when they are transparent and highly colored. However, few of the stones used for seals by the ancients were as brightly colored as *smaragdos*, and this may be the reason why the effect was noted only in the case of this one stone. It is possible, too, that Theophrastus may have based his statement on a single observation of *smaragdos* and that he did not attempt to experiment with other stones to see if they behaved in a similar way.

24. *It is also good for the eyes, and for this reason people carry seals made of it, so as to see better.*

The verb βλέπειν without a preposition following it does not mean "to look at" but "to see," and some adverb like εὖ ("well") is really needed here for clarity. The meaning must be "to see better" or "to improve their sight." Pliny[110] dwells at length on the pleasing green color of *smaragdus* and its supposed beneficial effect on the eyes. Though Theophrastus classifies *smaragdos* as one of the stones on which seals are engraved, Blümner[111] inferred from the wording of this particular passage that the allusion is to uncut ring stones. This conclusion is apparently supported by Pliny's statement[112] that it was forbidden to engrave the surface of *smaragdus*. Since Roman emeralds were rarely engraved, Pliny's statement seems to be confirmed, and it is probable that he is speaking of true emeralds at this point.[113] However, since no emeralds of purely Greek provenance have been found, it seems reasonably certain that Theophrastus is not alluding to emerald

---

[110] XXXVII, 62-63.
[111] Blümner, *Technologie und Terminologie der Gewerbe und Künste bei Griechen und Römern*, Vol. III, pp. 241-42.
[112] XXXVII, 64.
[113] King, *The Natural History of Precious Stones and of the Precious Metals*, p. 298.

here. The early Greek gems described by Furtwängler[114] which are green in color are all made of quartz, and it is very probable that here, and in the previous sections where Theophrastus says that *smaragdos* was used for seals, green quartz in the form of plasma or prase is to be understood.

24. *But it is rare and of small size, unless we are to believe the records about the Egyptian kings.*
Elsewhere in the treatise Theophrastus alludes to the scarcity or small size of the *smaragdos* (secs. 8, 26, 27, 34), and some writers[115] have concluded that emerald is meant whenever he alludes to the stone in this way. There is, however, no evidence that Theophrastus ever used the name *smaragdos* for an emerald. Moreover, the stones listed with it in section 8, which he describes as small and rare, are not especially rare and are not always small. In these respects they differ little from plasma or prase. Therefore, when Theophrastus refers to *smaragdos* in this way, he may well be speaking of the green quartz commonly used as a seal stone by the Greeks.

Although there was probably some real basis for the reports about *smaragdoi* of great size, Theophrastus shows clearly by his wording that he hesitated to accept these accounts without question. This was not because he thought that green stone objects of large dimensions did not exist, but because he doubted that the kind used for seals ever occurred in large masses. Certainly, if the accounts concerning them were not mere inventions or gross exaggerations, these large Egyptian *smaragdoi* could not have been actual precious stones. It is possible, however, that they may have been composed of malachite, which even now is sometimes regarded as a semiprecious ornamental stone. This native copper carbonate has been found in the form of solid blocks weighing several thousand pounds; in fact, it is the only bright green mineral substance that occurs in such large pieces. In modern times the copper mines in the Ural Mountains have been the source of some very large blocks of malachite. For example, the largest

[114] *Die antiken Gemmen*, II, 37-69, 152-53.
[115] E.g., Blümner, *Technologie und Terminologie der Gewerbe und Künste bei Griechen und Römern*, Vol. III, p. 240.

COMMENTARY

piece of flawless malachite said to have been found in the mines at Gumeshevsk weighed about 3,000 pounds, and in 1855 a mass was found at Nizhne-Tagilsk that weighed around 50,000 pounds, though this was of inferior quality.[116] It is quite possible that single pieces of malachite of huge size were also found in the early stages of working some of the great copper deposits of antiquity. That the mineral was available to the Egyptians is certain, since malachite was evidently the chief ore in the copper mines of the Sinai Peninsula, which were for centuries important sources of copper and copper minerals for Egypt.[117] The block of *smaragdos* (6 x 4½ ft.) that is said to have been sent to Egypt by a Babylonian king is as large as some of the modern objects made of polished malachite, such as the table tops, bathtubs, and panels for walls or columns which can be seen in certain European museums and other buildings, but it is improbable that malachite could have been the material of the four stones about sixty feet long mentioned as being placed in an obelisk. However, in the next section, where Theophrastus mentions a large pillar of green stone, he definitely suggests that this may have been composed of false *smaragdos*, a term that almost certainly denoted malachite. But this statement cannot provide a definite identification, since it is apparent that Theophrastus had no first-hand knowledge of the substance from which these large objects were made. The tradition that the malachite columns now in the church of Hagia Sophia at Constantinople originally came from the temple of Diana at Ephesos[118] suggests that large pillars of malachite actually existed in antiquity. It seems probable, therefore, that these Egyptian *smaragdoi* were composed, at least in part, of malachite, unless they were made of some common massive green rock such as serpentine, which is known to occur frequently in Egypt.[119] But there is a serious objection to identifying them as serpentine or some other green rock; for the term *smaragdos* was apparently applied only to mineral substances of a bright green color. That

[116] M. H. Bauer, *Edelsteinkunde* (Leipzig, 1932), p. 700. These weights are given in kilograms.
[117] Lucas, *Ancient Egyptian Materials and Industries*, pp. 231-35; Partington, *Origins and Development of Applied Chemistry*, pp. 60-63.
[118] Bauer, *Edelsteinkunde*, p. 701.
[119] Lucas, *Ancient Egyptian Materials and Industries*, pp. 479-80.

THEOPHRASTUS ON STONES

these huge *smaragdoi* were made of green glass, as some have suggested, is not very probable, because, as Partington[120] has pointed out, it is very doubtful from a technical standpoint that such enormous pieces of glass could have been successfully fashioned by ancient glassworkers. The existence of large *smaragdoi* in Egypt is mentioned by other ancient writers. Pliny,[121] for instance, after quoting the statements of Theophrastus almost word for word, mentions a recent example in the Labyrinth of Egypt; there Apion saw a colossal statue of Serapis, nine cubits in height, which was composed of *smaragdus*.

25. *tanoi.*

In the manuscripts and the Aldine edition, the first part of this word is missing and only the last four letters (ανῶν) have survived. Turnebus, who was followed by Hill, thought that the best emendation was τανῶν; the emendation of Furlanus, which Wimmer preferred, was βακτριανῶν. It is true that stones from Bactria are mentioned by Theophrastus in section 35, and Pliny[122] lists the Bactrian as a particular variety of *smaragdus*. But both Theophrastus and Pliny say that Bactrian stones are small, whereas the stone in question here was evidently a mineral substance found in pieces of considerable size. Moreover, as the notes on section 35 explain, it is highly probable that Pliny classed the Bactrian stones as one kind of *smaragdus* only because he misunderstood the meaning of Theophrastus in section 35. The evidence for the reading τανῶν is that Pliny[123] clearly lists *tanos* as a kind of *smaragdus*. He also adds that it came from Persia and was of an unsightly green color. His descriptive details seem to indicate some source other than the present passage of Theophrastus, unless the information given by Pliny was originally contained in this passage and dropped out later. This possibility is not at all unlikely, since almost all the other passages in this same section of Pliny's work are direct quotations from this part of the treatise. It is impossible to determine what was originally written in the lacuna before the letters ανῶν, and it is significant that De Laet made no

---

[120] *Origins and Development of Applied Chemistry*, p. 132.
[121] XXXVII, 74-75.   [122] XXXVII, 65.
[123] XXXVII, 74. The text reads: *Inseritur smaragdis et quae vocatur tanos e Persis veniens gemma, ingrate viridis atque intus sordida.*

change in the reading of the manuscripts, since he was not satisfied with either of the two proposed emendations. Stephanides has recently decided in favor of τανῶν, and this is certainly better than βακτριανῶν.

Though both the locality and the brief description of *tanos* given by Pliny apply to green turquoise, the large size of the stone mentioned by Theophrastus is definitely against this identification. Possibly *tanos* was the proper name for green turquoise, and the large slab at Tyre was actually composed of some other stone that only resembled this mineral. Theophrastus appears to have this possibility in mind when he suggests that the material of the slab may have been a "false *smaragdos*" rather than *tanos*. The former term was clearly used to denote malachite. Green turquoise and malachite, because of their similar appearance, were apparently often confused in ancient times, and this confusion is not entirely absent from the works of modern writers who have attempted to identify the green stones used by ancient peoples.[124]

25. Tyre.

Very likely Theophrastus obtained part of his information about this huge green stone from Herodotus,[125] who visited Tyre and saw this remarkable column or slab, which he describes briefly. Herodotus says that the column shone at night, and some commentators have suggested that it may have been composed of colored glass with a light inside it.[126] Though this seems an attractive explanation because of the reputed skill of the glassworkers of Phoenicia,[127] the descriptions apparently refer to a natural stone rather than to an artificial material. Furthermore, as was pointed out in the notes on section 24, it is highly improbable that ancient artisans could have fashioned any very large object out of glass. If it was composed of glass at all, this column must have been made of numerous small pieces fastened together in some way. On the whole, it seems much more probable that it was

[124] Lucas, *Ancient Egyptian Materials and Industries*, p. 457.
[125] II, 44.
[126] G. Rawlinson, *The History of Herodotus* (London, 1858-1860), Vol. II, pp. 81-82; W. W. How and J. Wells, *A Commentary on Herodotus* (Oxford, 1912), Vol. I, p. 188.
[127] Partington, *Origins and Development of Applied Chemistry*, pp. 454-55.

made of some natural material. Though this may have been malachite, serpentine is an attractive possibility, since not only is this green stone found in large blocks but some varieties are translucent enough to allow light to shine through the stone if it is in thin layers.

25. *Herakles.*
In reality, this was probably the Tyrian *Melkart*, the *Baal* of the Old Testament, whom the Greeks identified with Herakles.

25. *false smaragdos.*
Since this was found in copper mines in pieces of considerable size, it seems clear that it must have been malachite, the green basic copper carbonate. Probably the term was applied only to massive malachite, which was good enough to be used for ornamental purposes, since the earthy forms of this mineral were apparently included under the name *chrysokolla*. It is reasonably certain that the term "false *smaragdos*" was not applied to imitation green stones composed of glass or stained rock crystal.

25. *Chalcedon.*
All the manuscripts have the reading Καρχηδόνι (Carthage), and this was accepted by Hill and the editors who preceded him. Schneider and Wimmer have changed the name to Χαλκηδόνι (Chalcedon). This must be right, as there are no islands near the site of ancient Carthage that are known to have any copper minerals on them. On the other hand, at least one of the Prince Islands in the Sea of Marmora close to the shore of ancient Chalcedon is known to have been the site of ancient copper deposits. The author of *De Mirabilibus Auscultationibus*[128] names Δημόνησος (Demonesos) as an island of the Chalcedonians where the copper minerals *kyanos* and *chrysokolla* were found, as well as the copper that was used for making certain ancient statues. Pliny[129] lists *Demonnesus* as one of the islands in the Sea of Marmora at the entrance to the Bosporus. This island has been plausibly identified with the modern Khalki, upon which there are copper minerals and traces of ancient mining operations.[130]

[128] Sec. 58. [129] V, 151.
[130] Pauly-Wissowa, *Real-Encyclopädie*, III², p. 2093; V¹, p. 145.

COMMENTARY

26. *chrysokolla*.

This is mentioned again in sections 39, 40, and 51 as an ore or mineral found in mines. Though Theophrastus does not describe it anywhere, his repeated allusion to its occurrence in copper mines clearly indicates that it was a copper mineral. Later writers also mention that it occurred in gold mines, and some of them say that it was found in mines containing other metals; but this occurrence was evidently due to the presence of copper minerals in such mines, as is explicitly stated by Isidorus.[181] In section 39 *chrysokolla* is said to occur in native *kyanos*, which was azurite, the blue copper carbonate, showing clearly that *chrysokolla* in this case was malachite, the green copper carbonate.[182] The kind of *chrysokolla* mentioned in this treatise evidently corresponds to the natural kind of *chrysocolla* mentioned by Pliny[183] as an exudation or incrustation found in mines. Dioscorides[184] states that the best kind of *chrysokolla* was of a leek-green color. The descriptions given by ancient writers show that the name *chrysokolla* or *chrysocolla*, referring to a natural product, was given to any bright-green copper mineral that occurred as an earthy incrustation. From this it follows that the name must have denoted malachite, green copper carbonate, when it was in an earthy form, and also the amorphous green copper silicate which is still called chrysocolla at the present time. However, the descriptions of Pliny[185] indicate that *chrysocolla* as the name of a mineral was more often applied to malachite than to what is now called chrysocolla. From its peculiar name, which means "gold glue," some scholars[186] have erroneously concluded that the *chrysocolla* of the ancients was borax or some other soldering flux, though there is no basis for this conclusion other than the name and the stated use of the material. In the sixteenth, seventeenth, and eighteenth centuries borax was frequently called "chrysocolla," and this circumstance may have caused the wrong identification. The name was first given to borax by Agricola,[187] who may have misunderstood its meaning.

[181] XIX, 17, 10.   [182] See also the notes on sec. 39.   [183] XXXIII, 86.
[184] V, 104 (Wellmann ed., V, 89).   [185] XXXIII, 86.
[186] E.g., F. Hoefer, *Histoire de la Chimie* (Paris, 1866-1869), Vol. I, p. 173; Vol. II, p. 401; Lewis and Short, *Harper's Latin Dictionary*, *s.v.* But compare the *Thesaurus Linguae Latinae*, *s.v. Metallum* ("Malachit") *praecipue ad ferruminandum usurpatum*.
[187] G. Agricola, *De Natura Fossilium* (Basel, 1558), p. 206.

The method of using copper minerals for soldering gold in ancient times has evidently puzzled many modern writers on early technical arts, though the correct explanation was given at the end of the eighteenth century by Guettard, who showed experimentally that malachite, for example, could be used as a solder for gold.[138] All that is required is that some suitable reducing agent, such as charcoal or organic matter, should be present, and that the temperature should be high enough to reduce the mineral to copper and to make this melt and alloy with the gold. It has recently been shown for the first time that certain types of ancient goldwork could only have been produced by a soldering process of this sort. Some Etruscan and Greek works of art contain delicate patterns formed by minute grains of gold or very fine wire joined to a background of solid metal; it has been found by experiment that they could not have been joined together by the direct application of solder in the form of molten metal or by any process of fusion welding. Furthermore, microscopic examination of examples of ancient goldwork has shown the use of this reduction method of soldering.[139]

It is interesting to note that this ancient method of soldering gold by the reduction of a copper compound *in situ* is the subject of a modern patent[140] issued to a Mr. Littledale, who may be regarded as the rediscoverer of an old method of soldering which is used for goldwork of great delicacy.

The name *chrysocolla* was applied by writers later than Theophrastus to artificial copper preparations used for soldering gold. Pliny[141] describes a mixture of this sort. In addition, he mentions under *chrysocolla* a preparation that contained gold and silver in addition to copper salts. By further extension of the original meaning, the name was applied to alloys used for the soldering of gold. In the *Leyden Papyrus X* are two recipes (Nos. 31 and 33) for the preparation of such gold solders.[142] In Number 31, the alloy of copper, silver, and gold is called τὸ χρυσόκολλον.

---

[138] Bailey, *The Elder Pliny's Chapters on Chemical Subjects*, Part I, p. 206.
[139] H. Maryon, *Technical Studies in the Field of the Fine Arts*, V (1936), 88-95.
[140] British patent, 1934, No. 415181.
[141] XXXIII, 93.
[142] M. Berthelot, *Archéologie et Histoire des Sciences* (Paris, 1906), pp. 280, 282.

COMMENTARY

For remarks on the use of *chrysokolla* as a pigment, see the notes on section 51.

27. *iaspis*.

Though the word "jasper" is derived from the Greek ἴασπις and is often used to translate the Greek word and its Latin equivalent *iaspis*, the one fact that is most certain about the ancient name is that it did not designate the kinds of opaque colored silica that are now called jasper. The descriptions of ancient writers usually show that the name denoted certain transparent or translucent stones, and there is no definite evidence that it was ever applied to an opaque mineral substance. Though Theophrastus does not allude clearly to *iaspis* as a transparent stone, there is no such uncertainty about the descriptions left us by other ancient writers. Pliny opens his account of the stone with the words: *Viret et saepe tralucet iaspis*[143] (*iaspis* is green and often translucent). Later he[144] mentions a kind that resembles rock crystal. Pliny[145] also alludes to imitations of *iaspis* made of glass. Moreover, the descriptions of Dioscorides[146] show that the name was not applied to an opaque stone. Dionysius Periegetes describes it as being watery,[147] green and translucent,[148] and cloudy.[149] This is not an appropriate description of the stone that is now called jasper. Though there can be no doubt that it was not our modern jasper, there is less certainty about its positive identification.

Since Theophrastus shows its relationship to *smaragdos* in this passage, one might infer that *iaspis* was a green stone, and this color is mentioned by all ancient writers who describe it. Indeed, some of them mention this color only, and in the *Stockholm Papyrus*,[150] where a recipe is given for the preparation of an artificial *iaspis*, it is clear from the ingredients that the resulting product was a green stone. On the other hand, Pliny[151] refers to *iaspis* of various other colors, such as blue and rose, as well as to a colorless variety. In fact, he classifies the best as having a shade of purple and assigns only third place to the green kind. He also

[143] XXXVII, 115.   [144] XXXVII, 116.   [145] XXXVII, 117.
[146] V, 159 (Wellmann ed., V, 142).   [147] 782 (ὑδατόεσσαν).
[148] 1120 (χλωρὰ διαυγάζουσαν, lit., "shining through green").
[149] 724 (ἠερόεσσαν).   [150] Lagercrantz, *Papyrus Graecus Holmiensis*, p. 15.
[151] XXXVII, 115-16.

mentions smoky and turbid kinds. Dioscorides[152] gives a similar but less extensive list of the varieties of this stone. The ancient descriptions seem to show that *iaspis* was a generic term that usually denoted those varieties of transparent or translucent quartz to which special names such as *sardion* or *crystallos* were not applied. Thus the green kind was probably plasma or chrysoprase, the smoke-colored kind was smoky crystalline quartz or smoky chalcedony, the rose-colored kind was rose quartz, and the blue kind was common blue chalcedony. All these varieties of quartz were used as materials for ancient engraved stones. It is significant that Pliny includes *sphragis* or seal stone under the term *iaspis*. This suggests that *iaspis* was a name applied to some varieties of chalcedonic or clear quartz used for seals.

It seems likely, however, that other minerals besides quartz which were similar in appearance were included under the ancient name. Thus it has been suggested that jade or nephrite was called *iaspis* in antiquity.[153] Certainly the kind of *iaspis* mentioned by Pliny,[154] who describes it as a green stone with one or more white lines running through it, would seem to correspond to jade or nephrite. Pliny implies that this stone was used as an amulet, and had its origin in the East, and both these clues tend to support this particular identification. In the same way, still other minerals such as fluorite, which in some of its forms resembles certain varieties of colored quartz, may have been classified under *iaspis* in ancient times.

Theophrastus shows by his remarks in section 37 that certain kinds of true jasper used in antiquity were given particular names. See also the notes on that section.

27. *It is said that a stone was once found in Cyprus half of which was smaragdos and half iaspis, as if it had not been entirely changed from the watery state.*

This passage is quoted by Pliny,[155] who has clearly obtained his information from Theophrastus. His wording is as follows: *et in*

---

[152] V, 159 (Wellmann ed., V, 142).
[153] Moore, *Ancient Mineralogy*, p. 219; J. Berendes, *Des Pedanios Dioskurides aus Anazarbos Arzneimittellehre* (Stuttgart, 1902), p. 551.
[154] XXXVII, 118.   [155] XXXVII, 75.

· 108 ·

COMMENTARY

Cypro inventum ex dimidia parte smaragdum, ex dimidia iaspidem, nondum umore in totum transfigurato. Here the Latin *umore* reproduces the Greek τοῦ ὕδατος, which can be translated as "the watery state" or "its watery state." The sense of the passage seems to be that a piece of stone or a crystal was once found in Cyprus half of which had the green color of *smaragdos*, while the other half had a limpid or watery appearance and was probably colorless or only slightly colored. This could have been green quartz in a matrix of clear colorless quartz or colorless chalcedony, though it seems rather more likely that the allusion is to a crystal. Crystals of this sort are not uncommon. Tourmaline frequently occurs in the form of transparent crystals that sometimes are green at one end and have a different color or are colorless at the other end. It is easy to see how mineral occurrences of this kind would lead to the idea that one sort of precious stone could originate from another. Many later writers on mineralogy advance this idea, though Theophrastus was the first to express it.

28. *lyngourion.*

This substance is also mentioned by other ancient authors such as Strabo,[156] Dioscorides,[157] and Pliny.[158] Though many commentators have tried to identify it, unfortunately they have disagreed in their conclusions. Some have thought that it was a fossilized animal substance, others that it was a particular kind of precious or semiprecious stone, others that it was amber, and still others that it was a kind of fossil resin resembling amber.

Some early modern writers on mineralogy identified this substance as belemnite, fossil cuttlefish bone. De Boodt[159] gives *lyncurius* as a synonym for *belemnite*, and later writers such as Woodward[160] definitely identify the *lyngourion* or *lyncurium* of the ancients in this way. However, as Hill,[161] Watson,[162] and Beckmann[163] have clearly pointed out, it could not possibly have been

[156] IV, 6, 2.   [157] II, 100 (Wellmann ed., II, 81, 3).   [158] XXXVII, 52.
[159] A. B. De Boodt, *Gemmarum et Lapidum Historia* (Leyden, 1647), p. 476.
[160] Cited by Hill, *Theophrastus's History of Stones*, p. 73.
[161] *Ibid.*, p. 74.
[162] W. Watson, *Philosophical Transactions of the Royal Society of London*, LI (1759), 396.
[163] J. Beckmann, *History of Inventions, Discoveries, and Origins* (London, 1846), Vol. I, p. 86.

· 109 ·

belemnite, since this fossil substance does not have the hardness, the transparency, or the electrostatic-attractive properties ascribed to *lyngourion* or *lyncurium*.

The *lyngourion* of Theophrastus has been most often identified with a particular kind of precious or semiprecious stone. De Laet,[164] who referred to an earlier suggestion of Epiphanius, says that "the description of *lyncurium* is certainly not inappropriate to the hyacinth of modern writers." Hill[165] rightly rejects its earlier identification with belemnite, but, ignoring the possibility that it was amber, he decides with De Laet that it was "hyacinth." Apparently he uses this word to describe certain varieties of garnet. Watson[166] identifies the *lyngourion* of Theophrastus with tourmaline, but evidently his opinion is partly based on the attractive properties of heated tourmaline which had recently been discovered. This identification is repeated by various later writers. For example, Dana[167] states that *lyncurium* is supposed to be the ancient name for common tourmaline. However, the absence of tourmaline among surviving examples of ancient gems is clearly against this view. Its identification as red garnet or red tourmaline may have been based to some extent on the emendation πυρρά ("flame-colored") which was substituted by Furlanus for the manuscript reading ψυχρά ("cold") and was adopted by several editors, including Hill. Wimmer has the manuscript reading in the neuter form ψυχρόν. Though the emendation seems to agree better with the description of its color (*"colorem igneum"*) given by Pliny,[168] who apparently derived his information from this account of Theophrastus, the manuscript reading is more suitable, since it is evident from section 31 that the color of *lyngourion* was yellow. The suggestion has been made that it was the stone known today as the hyacinth or jacinth,[169] but this seems to be partly due to a confusion of mineralogical names. That *lyngourion* could have been any of the gem varieties of zircon is highly improbable; no

---

[164] J. De Laet, *De Gemmis et Lapidibus Libri Duo* (Leyden, 1647), p. 155. The text reads: *sane descriptio lyncurii non male convenit cum hyacintho Neotericorum*.
[165] *Theophrastus's History of Stones*, pp. 73-77.
[166] *Philosophical Transactions of the Royal Society of London*, LI, 396.
[167] J. D. Dana, *Manual of Mineralogy and Petrography* (New York, 1909), p. 306.
[168] XXXVII, 53.
[169] See *Encyclopaedia Britannica* (14th ed.), s.v. *hyacinth*.

COMMENTARY

ancient gems of zircon have been found, and since they occur principally in Ceylon, it is unlikely that the ancient Greeks could have been acquainted with them.

Though the hardness and transparency ascribed to *lyngourion*, as well as its use for seals, suggest that it was a kind of precious or semiprecious stone, certain of its qualities, as described by Theophrastus, show that it was not an inorganic material at all. In particular the phrase καθάπερ λίθος ("like stone"), used with reference to its hardness, definitely indicates that it was not a stone, and the electrostatic properties ascribed to it, though they could apply to a polished gem, point to a more easily electrified substance such as amber, which is of vegetable origin.

The explicit statements of Strabo, Dioscorides, and Pliny on the nature of this substance have not been sufficiently considered in many of the attempts to identify it. In his discussion of the territory of the Ligurians, Strabo[170] remarks: "The *lingourion*,[171] too, is plentiful in their country, and some call this amber." Dioscorides says that "the urine of the lynx, which is called *lyngourion*,[172] is believed to be transformed into a stone as soon as it is voided, and so it has a foolish story connected with it; for this is what some people call the amber that attracts feathers . . . ."[173] In his discussion of the various names given to amber, Pliny says:

"*Demostratus lyncurium vocat . . . alios id dicere langurium . . .*"[174] (Demostratus calls it *lyncurium* . . . ; others call it *langurium* . . .). After discussing the varieties, properties, and uses of amber, Pliny adds this statement: "*De lyncurio proxime dici cogit auctorum pertinacia, quippe, etiamsi non electrum id, tamen gemmam esse contendunt, fieri autem ex urina quidem lyncis, sed et genere terrae, protinus eo animali urinam operiente, quoniam invideat homini, ibique lapidescere. esse autem, qualem in sucinis, colorem igneum, scalpique nec folia tantum aut stramenta ad se rapere, sed aeris etiam ac ferri lamnas, quod Diocli cuidam Theophrastus quoque credit. ego falsum id totum arbitror nec visam in aevo nostro gemmam ullam ea appellatione.*"[175] (The obstinacy of authors compels me to speak next of *lyncurium*, for even though they state that it is not amber but a precious stone, yet they assure us that it is formed from the urine of the lynx, though it also contains a kind

[170] IV, 6, 2.
[171] λιγγούριον.
[172] λυγγούριον.
[173] II, 100 (Wellmann ed., II, 81, 3).
[174] XXXVII, 34.
[175] XXXVII, 52-53.

· III ·

of earth; the animal immediately covers up his urine because of his envy of mankind, and there it turns into stone. Moreover, like amber it has the color of fire, can be engraved, and attracts to itself not only leaves or straws, but even thin pieces of bronze and iron, as Theophrastus believes on the authority of a certain Diocles. I consider that all this is untrue and that a precious stone of this name has not been seen in our time.)

From these statements it seems likely that the substance variously known as *lyngourion, lingourion, lyncurium*, or *langurium* was none other than amber, and this identity has been upheld by some commentators.[176] However, if the substance which Theophrastus calls *lyngourion* was really identical with the one called *electron*, it remains to be explained why he discusses them as though they were different substances. Possibly he was unaware that the same substance was known under these two different names. Certainly his statements about the mode of origin of these two substances indicate a lack of first-hand information that might easily have led him to just such a confusion of names. On the other hand, when he says that *lyngourion* had a high degree of transparency, he does suggest that there may have been a real difference between this substance and the one called *electron*, since the striated or clouded varieties of amber could not be so characterized. Possibly the name *lyngourion* was a special one applied only to flawless varieties of amber valued for purposes of adornment, whereas *electron* was the general name for amber, or the name applied to the less valued varieties.

Finally there is the theory[177] that *lyngourion* was not genuine amber but some sort of fossil resin either allied to amber or resembling it, but this seems to be based on what Theophrastus says about its mode of origin. His statements certainly indicate that *lyngourion* was found only in the ground, which is not true of genuine amber. However, it might also be inferred from what he says in section 29 that *electron* was also found in the ground and was therefore not genuine amber. Hence his statements give no valid reason for regarding *lyngourion* as different from amber. Actually, of course, the fanciful tale he tells about the formation

---

[176] Blümner, *Technologie und Terminologie der Gewerbe und Künste bei Griechen und Römern*, Vol. II, pp. 381-82 (footnote).
[177] *Ibid.*, p. 382.

of *lyngourion* shows that he personally knew nothing about this. That *lyngourion* was not one of the soft fossil resins seems evident from the remark he makes about its hardness.

On the whole, therefore, *lyngourion* can be identified with reasonable certainty as either amber or some particular variety of amber.

28. *bits of wood.*

In quoting this passage about the attractive powers attributed to *lyngourion*, Pliny[178] mentions leaves ("*folia*") rather than wood. For this reason Wimmer added the word φύλλα ("leaves") in brackets after ξύλον ("wood"), indicating that the text may have originally contained this word. Since Pliny is not always accurate in his quotations, this is far from certain. However, the context makes it clear that the word ξύλον should be understood to mean bits of wood or shavings.

Theophrastus does not say that it is necessary to rub *lyngourion* in order to induce its attractive power, but other ancient authors also fail to mention this when they speak of the electrostatic attraction that amber displays. It is uncertain whether Theophrastus was even aware of the necessary part played by friction in producing this phenomenon, especially as his statements show that he was dependent upon others for his information.

28. *Diokles.*

This is generally considered to be the earliest mention of Diokles of Karystos, a famous Greek physician and writer of the fourth century B.C. who in all likelihood was for some time a contemporary of Theophrastus at Athens. The significance of the allusion to Diokles for dating this treatise of Theophrastus is discussed in the notes on section 59.

28. *it is better when it comes from wild animals rather than tame ones and from males rather than females; for there is a difference in their food, in the exercise they take or fail to take, and in general in the nature of their bodies, so that one is drier and the other more moist. Those who*

[178] XXXVII, 53.

are experienced find the stone by digging it up; for when the animal makes water, it conceals this by heaping earth on top.

Theophrastus appears to be the first to relate this curious story, but whether he was the actual author of it is uncertain. He may have depended, as for some of his other information about *lyngourion*, on the statements of other writers, or, as is even more probable, he may be repeating a popular tale that was widely known and believed. This story undoubtedly arose from the name of the substance (lynx-urine), though its color and general appearance may also have been a factor in the origin of the story. It is not unlikely that someone tried to invent an etymology for the name after its original pronunciation and spelling had been corrupted and its real origin forgotten. Though Theophrastus fails to state explicitly in this treatise what animal was supposed to produce *lyngourion*, the lynx is specifically named in all later accounts. Thus, for example, the animal is so named in the accounts of Dioscorides and of Pliny that have already been quoted. The animal is also named by Pliny in other passages[179] dealing with the subject, and in a fragment quoted by Photius[180] Theophrastus specifically names the lynx as the animal whose urine was utilized for seals. Theophrastus does not say anywhere how the liquid urine was transformed into the solid stony substance, but Pliny is explicit on this point. In one place[181] he states that the urine either congealed or dried, and in another place,[182] where he seems to be depending more on the opinions of others, he says that it was the urine of the lynx and a kind of earth that hardened together to form the stone. Ovid[183] remarks that the urine hardened on contact with air. It is clear from the variety of the explanations given by later authors, and even more from the other variations in the details of the story, that it must have been widely known, and lost nothing from being retold. The following passage from Pliny shows how many alterations were introduced into the story as it was passed along:

[179] VIII, 137; XXXVII, 34.
[180] *Bibliotheca* (Bekker ed.), p. 528, col. 2.
[181] VIII, 137.   [182] XXXVII, 52.
[183] *Metamorphoses*, XV, 415.

COMMENTARY

"Demostratus lyncurium vocat et fieri ex urina lyncum bestiarum, e maribus fulvum et igneum, e feminis languidius atque candidum; alios id dicere langurium et esse in Italia bestias languros. Zenothemis langas vocat easdem et circa Padum iis vitam adsignat, Sudines arborem, quae gignat in Liguria, vocari lynca. in eadem sententia et Metrodorus fuit."[184] (Demostratus calls it *lyncurium* and says that it originates from the urine of the animal known as the lynx, that of the male being reddish and fiery, that of the female rather pale and even white; others call it *langurium*, there being in Italy animals known as *languri*. Zenothemis calls them *langae*, and assigns the region of the Po River as their habitat. Sudines says that the tree which produces it in Liguria is called the *lynx*. Metrodorus also was of the same opinion.)

Perhaps the most curious elaboration of the story was the belief that the lynx hid its urine because it did not wish men to possess the valuable stone formed from it.[185] Pliny mentions this strange belief in two passages,[186] one of which has already been quoted. It is also mentioned in section 76 of the pseudo-Aristotelian work *De Mirabilibus Auscultationibus*. The passage reads as follows: καὶ τὴν λύγκα δέ φασι τὸ οὖρον κατακαλύπτειν διὰ τὸ πρὸς ἄλλα τε χρήσιμον εἶναι καὶ τὰς σφραγῖδας. (They say that the lynx also covers up its urine because it is useful for seals and other purposes.) Though Theophrastus says nothing about this curious belief in his account of the story in this treatise, it must have been current in his day, since he takes the trouble to refute it in the fragment quoted by Photius, which contains various examples of envy or jealousy felt by animals toward men. The pertinent passage reads as follows: καὶ ἡ λὺγξ κατακρύπτει τὸ οὖρον ὅτι πρὸς τὰς σφραγῖδας καὶ πρὸς ἄλλας χρείας ἐπιτήδειον. ἀλλ' ὅτι μὲν οὐ διὰ φθόνον ταῦτα ποιεῖ τὰ ζῷα ἀλλ' οἱ ἄνθρωποι ἐκ τῆς ἰδίας ὑπολήψεως ταύτην αὐτοῖς περιῆψαν τὴν αἰτίαν παντὶ δῆλον.[187] (And the lynx conceals the urine because it is suitable for seals and other purposes. But it is obvious to everyone that animals do not do this from envy, but that men have brought this charge against them because of their own prejudice.) This passage, though confirming the belief of Theophrastus that *lyngourion* was formed

[184] XXXVII, 34.
[185] Plutarch (*Moralia*, 962F, *De sollertia animalium*) seems to hint at this belief, though he merely says that lynxes conceal *lyngourion*.
[186] VIII, 137; XXXVII, 52.
[187] Text of Wimmer, *Theophrasti Eresii Opera* (Teubner ed.), III, fr. CLXXV, 10-14.

· 115 ·

from the urine of the lynx, also indicates that the credulity of Theophrastus had its limits. Moreover, it seems to show that in the treatise *On Stones* Theophrastus may have purposely omitted the part of the story that accuses the lynx of hiding its urine owing to envy or jealousy. It is not unlikely that this fanciful explanation was generally considered an essential part of the story in his day.

28. *This stone needs working even more than the other kind.*

This statement has been taken as evidence that *lyngourion* was a very hard stone that was difficult to cut or grind, though this difficulty is certainly not encountered in working amber. Whether this statement is true is by no means certain, especially when it is compared with the statements that precede it; but if it is assumed that it is true or that it contains an element of truth, it may refer only to the time needed for polishing the material when it is cut. If this interpretation is accepted, the statement could refer to amber as well as to a hard stone, so that it cannot be taken as evidence that *lyngourion* was not amber or a variety of amber.

29. *And since amber is also a stone—for the kind that is dug up is found in Liguria—the power of attraction would belong to this too.*

Theophrastus has previously mentioned Liguria as a source of amber in section 16. The boundaries of ancient Liguria were not well defined; though the territory was probably restricted largely to northwestern Italy at the time of Theophrastus, it extended far along the coast of southern Gaul at an earlier date.[188] At no time, however, did it include regions where amber occurred, nor was amber found in any region near it at the time of its greatest extent. The truth appears to be that amber was brought from the coasts of the North and Baltic seas by trade routes through Gaul, and that the Ligurians acted only as traders in this product.[189] Theophrastus, like all other classical Greek writers who touch on the subject, apparently knew nothing about the real nature or the real source of amber.

[188] W. W. Hyde, *Roman Alpine Routes* (Philadelphia, 1935), pp. 43, 134.
[189] *Ibid.*, pp. 42-43.

COMMENTARY

Though Thales of Miletus is generally regarded as the first to mention that amber has the property of attracting light particles when it has been electrified by friction, his claim to this distinction actually rests on very uncertain grounds. That Thales was the first to mention this property can be inferred only indirectly from the following statement of Diogenes Laertius in his discussion of Thales: Ἀριστοτέλης δὲ καὶ Ἱππίας φασὶν αὐτὸν καὶ τοῖς ἀψύχοις μεταδιδόναι ψυχάς, τεκμαιρόμενον ἐκ τῆς λίθου τῆς μαγνήτιδος καὶ τοῦ ἠλέκτρου.[190] (Aristotle and Hippias say that, judging by the behavior of the lodestone and amber, he also attributed souls to lifeless things.) What Aristotle really says about this opinion of Thales is as follows: ἔοικε δὲ καὶ Θαλῆς ἐξ ὧν ἀπομνημονεύουσι κινητικόν τι τὴν ψυχὴν ὑπολαβεῖν, εἴπερ τὸν λίθον ἔφη ψυχὴν ἔχειν ὅτι τὸν σίδηρον κινεῖ.[191] (According to the reports made about him, Thales also seems to regard soul as a motive force, if indeed he said that the lodestone has a soul because it moves iron.) In other words, Aristotle, whom Diogenes Laertius quotes, does not even mention amber in his corresponding statement about Thales. Of course it may be inferred from these two statements that it was Hippias who said that Thales understood the attractive property of amber, but there is no way of confirming such an inference because the works of Hippias are not extant. It may even be that the allusion to amber in the statement of Diogenes Laertius is the result of a late interpolation, as has been suggested by Rossignol.[192] The first clear indication that the ancients knew about the attractive property of amber is given by Plato, who very briefly alludes to it in his *Timaeus*,[193] though he denies that it had a *real* power of attraction. The various statements of Theophrastus in sections 28 and 29 are certainly the earliest account of the properties of amber.

29. *The stone that attracts iron is the most remarkable and conspicuous example. This also is rare and occurs in few places.*

Though Theophrastus does not give a specific name to the lode-

[190] I, 1, 24.   [191] *De Anima* I, 2, 405A.
[192] J. P. Rossignol, *Les Métaux dans l'Antiquité* (Paris, 1863), p. 348.
[193] 80C.

· 117 ·

stone in this passage, he has in a preceding passage (sec. 4) apparently designated it as λίθος Ἡρακλεία ("Heraclean stone"). This seems to have been the common early Greek name for the lodestone, since Plato in one place[194] specifically states that this was what most people called it, and in two other places[195] he uses this as the name of the lodestone without special comment. However, the lodestone was frequently mentioned without the use of a special name, as in the present passage and in the passage from Aristotle that has just been quoted. It was sometimes described simply as "the stone," without any explanation that it was the one that attracted iron. For example, Theophrastus so designated it in the following passage in his *History of Plants*, where he is referring to certain plants that affect lifeless objects: τὰς δὲ καὶ ἕλκειν, ὥσπερ ἡ λίθος καὶ τὸ ἤλεκτρον.[196] (And some also have the power of attraction, like the stone [*sc.* the lodestone] and amber.)

Since the Greek world was so small and a very small area of the earth had been explored for minerals at that time, this statement of Theophrastus about the scarcity of the lodestone is undoubtedly correct for his day. Even at the present time specimens of magnetite that are actively magnetic are not very common in spite of the abundance of the mineral itself.

Although the statement in the present passage is a very early allusion to the phenomenon of magnetism, it is by no means the earliest that is known. As was indicated before, Thales of Miletus was probably the first to allude to this phenomenon, but the earliest direct statements about the lodestone and its special property are those of Plato. The passages in Plato's *Ion* constitute a particularly vivid description of the way the lodestone attracts iron. Though Plato in the *Timaeus* mentions amber and the lodestone together because of their attractive power, he does not suggest any particular connection between the properties of the two substances. Theophrastus is apparently the first to suggest explicitly that the lodestone should be classified with substances, such as amber, which exhibit the property of electrostatic attraction. Thus he may perhaps be considered the first to hint at a

[194] *Ion*, 533D. [195] *Ion*, 535E and *Timaeus*, 80C.
[196] IX, 18, 2.

possible connection between what we now call electricity and magnetism.

30. *hyaloeides.*

The name suggests some sort of glasslike stone, but it is mentioned by no other ancient author, and the very brief description given by Theophrastus is inadequate for certain identification. Various conjectures have been made by commentators. Hill,[197] for example, supposed that it corresponded to the *astrion* of Pliny,[198] whereas Werner[199] suggested that it might have been moonstone, and Lenz[200] that it might have been a natural glass. Stephanides[201] believed that various reflecting and transparent stones might have been known by this name, particularly the *lapis specularis* of Pliny,[202] which apparently included mica and selenite. Though it is quite possible that various materials of a glassy nature received the name ὑαλοειδής, it seems unlikely that such soft minerals as these were included, for Theophrastus is speaking of a stone or igneous material upon which seals were engraved. The real objection to identifying it as one of the glassy minerals is that, with the important exception of the various forms of quartz, all of which appear to have been known by their own special names, practically no specimens of engraved gems executed in such minerals have come down to us. Therefore, it seems not unlikely that the name may have been given to the various glass pastes that were by no means uncommon as a material for seals at the time of Theophrastus. About ten per cent of the engraved gems of the Hellenistic and early Roman period that are listed by Furtwängler[203] were executed in glass pastes of various colors. One objection to this identification is that Theophrastus would not be likely to list an artificial product like glass among stones of natural origin.

[197] *Theophrastus's History of Stones,* p. 80.
[198] XXXVII, 132.
[199] Cited by Moore, *Ancient Mineralogy,* p. 227.
[200] *Mineralogie der alten Griechen und Römer,* p. 21. The specific identification given by this commentator is "bouteillenstein," or bottle stone, a peculiar green natural glass also called moldavite or pseudochrysolite.
[201] *The Mineralogy of Theophrastus,* pp. 123-24.
[202] XXXVI, 160-62.
[203] Furtwängler, *Die antiken Gemmen,* Vol. II, pp. 130-46, 153-73. This figure and similar estimates mentioned elsewhere in this commentary are not given by Furtwängler but are based on his descriptions of antique gems.

However, his statements in section 48 of the treatise, where finished ceramic ware is described as stone, suggest that he would not have made any such distinction in classification. Judging from the reference Theophrastus makes to glass and vitreous earth in section 49, it is doubtful that he was aware of the artificial origin of glass; and even if he understood its mode of origin, it does not follow that he would have recognized that seals executed in paste were not made of a natural material, for since they were engraved, they were very different in appearance from large objects made out of glass by other methods.

30. *anthrakion.*
This is mentioned again in section 33. See the notes on that section.

30. *omphax.*
The Greek word ὄμφαξ usually means an unripe grape or some other unripe fruit, but it was often used in a metaphorical sense.[204] Perhaps Theophrastus is using it in this way here, as he appears to be the only author to give ὄμφαξ as the name of a stone. Galen[205] lists a stone with a similar name called ὀμφατῖτις (*omphatitis*), but, like Theophrastus, he gives no description. The name suggests that the stone resembled an unripe grape, i.e., it was green in color and in its natural state botryoidal. Furthermore, it is probable that it was either transparent or translucent, since the other stones discussed in this part of the treatise were not opaque. Practically the only mineral suitable for engraving that fulfills these conditions is prehnite, a hydrous calcium aluminum silicate usually colored apple-green by impurities and occurring normally in botryoidal masses lining cavities in igneous rocks. Prehnite is capable of receiving a high polish and has been used in modern times for inlaid work and ornaments.[206] However, ancient engraved stones of prehnite have apparently not been found, and though this may seem to bring the identification into question, it should be noted that surviving examples of ancient non-opaque green stones are scarce,

---

[204] Liddell-Scott-Jones, *Greek-English Lexicon*, s.v.
[205] *De simplicium medicamentorum temperamentis ac facultatibus*, IX (Kühn ed., XII, 207).
[206] Dana, *Manual of Mineralogy and Petrography*, pp. 317-18.

COMMENTARY

even though early authors speak of such stones as though they were common in their day. Possibly the stone was called *omphax* only because of its color, and for this reason it may have been some other stone such as green quartz. Stephanides[207] suggested that it might have been the variety of quartz now known as chrysoprase, a translucent form of chalcedony colored bright green by nickel. Here again, surviving examples of seals made of this material which would confirm the identification are lacking. Both these identifications should be regarded as conjectural, since the total lack of description makes it impossible to determine which particular stone was called *omphax*.

30. *rock crystal.*

This is the generally accepted identification of κρύσταλλος, and there can be little question that it is correct. In discussing *crystallus*, Pliny[208] refers to the Greek name as a word meaning a kind of ice; he clearly describes the hexagonal form of crystalline quartz, and even mentions the occurrence of variations in the characteristic pyramidal terminations of the crystals. Though Theophrastus lists it as one of the stones upon which seals were engraved, the small number of examples contained in modern collections of ancient engraved stones indicates that it was not commonly used for this purpose in his day. However, numerous specimens of engraved rock crystal of earlier date are known. About five per cent of the carved Mycenaean stones described by Furtwängler[209] and less than five per cent of the early Greek gems are made of this material, whereas less than one per cent of the stones ascribed to the Hellenistic and early Roman period are made of it. In Roman times rock crystal was commonly carved into relatively large objects such as dishes and drinking glasses, and these seem to have been highly valued.[210]

30. *amethyst.*

There is little doubt that the stone named ἀμέθυσον by Theophrastus was identical with our amethyst, a purple variety of

[207] *The Mineralogy of Theophrastus*, p. 76.
[208] XXXVII, 23.
[209] Furtwängler, *Die antiken Gemmen*, Vol. II, pp. 7-18, 25-27, 37-69, 130-46, 152-53, 153-73.
[210] Pliny, XXXVII, 27-29.

quartz. He mentions rock crystal and amethyst in one phrase as though they belonged together, stating further that they were found when dividing other stones. Colorless quartz and amethyst are often found in this way, and sometimes they are found together, in veins or cavities of massive rocks. They are also found lining the interior of geodes, and are revealed when these stones are broken. In the next section Theophrastus notes that ἀμέθυσον was wine-colored. These meager characterizations of amethyst are fortunately supplemented and confirmed by later writers. Thus, Pliny speaks of *amethystus* in this way: *causam nominis adferunt quod usque ad vini colorem accedens, priusquam eum degustet, in violam desinat fulgor.*[211] (The reason for its name is said to be that it approaches the color of wine, but before it reaches this color it shades off into violet.) In speaking of the varieties of amethyst, Pliny also says this: *quintum ad vicina crystalli descendit albicante purpurae defectu*[212] (a fifth kind approaches rock crystal very closely, the purple gradually fading off into white). The second quotation accurately describes the common kind of amethystine quartz which is only slightly or partially colored. Though it is likely that the same name would have been applied by the ancients to purple sapphire or purple fluorite because they are similar in color, there are no stones of this kind in collections of ancient engraved gems; this indicates that amethyst, if not the only purple stone known to the ancients, was at least the only one of this color that was engraved. About three per cent of the engraved gems of the Hellenistic and early Roman period that are catalogued by Furtwängler[213] are made of amethyst.

30. *sardion.*

The brief statements in this passage, when supplemented by what Pliny[214] says about *sarda*, show that σάρδιον was a generic name applied to those varieties of red chalcedony suitable for seals. These stones are simply quartz colored with small amounts of ferric oxide. Theophrastus plainly distinguishes two varieties: the one which is described as translucent but redder than the other

---
[211] XXXVII, 121.   [212] XXXVII, 123.
[213] Furtwängler, *Die antiken Gemmen*, Vol. II, pp. 130-46, 153-73.
[214] XXXVII, 105-107.

was apparently identical with what is now called carnelian, a bright red chalcedony of clear rich tint; whereas the darker kind probably corresponded to what is now called sard, usually a deep brownish-red chalcedony that becomes blood-red in color when light shines through it, though the color of the stone is sometimes so intense that it approaches black. It is interesting to note that both Theophrastus and modern writers on mineralogy distinguish only two varieties of red chalcedony suitable for gems, the difference being that a single name, suitably qualified, was used in his day, whereas two distinct terms are now employed.

The ancients made abundant use of red chalcedony for seals, as is shown by the large number of engraved stones of this material that have survived to the present day, and apparently red chalcedony was more often engraved by the peoples of the Aegean region than any other kind of precious or semiprecious stone. About twenty per cent of the Mycenaean engraved stones that are listed by Furtwängler,[215] thirty-five per cent of the early Greek stones, and thirty-five per cent of the Hellenistic and early Roman stones are made of red chalcedony. This early preference for red chalcedony is noted by Pliny in these words: *nec fuit alia gemma apud antiquos usu frequentior*[216] (among the ancients there was no precious stone in more common use). Carnelian was apparently the variety generally preferred in the earliest periods, and sard came into widespread use only in the Hellenistic period. About forty per cent of the stones made of red chalcedony that are listed by Furtwängler as Hellenistic are classified as sard, whereas less than five per cent of such stones that belong to the early Greek period are classified in this way. Possibly the exhaustion of the old sources of red chalcedony or the discovery of new sources would account for this variation in distribution, though it may have been simply the result of a change in fashion.

30. *And there are others, as has been mentioned before, which differ from one another, though they have the same name.*

No previous statement of this kind occurs in the treatise as it is

---

[215] Furtwängler, *Die antiken Gemmen*, Vol. II, pp. 7-18, 25-27, 37-69, 130-46, 152-53, 153-73.
[216] XXXVII, 106.

now known. However, at the beginning of section 23, where the differences in stones used for seals are mentioned in a general way, it is possible that something is missing that was contained in the original form of the treatise. For this reason Schneider proposed an emendation in section 23, which is listed in the critical notes on the text.

31. *And it is the same with the varieties of the lyngourion, for the female is more transparent and yellow than the other. Also, one kind of kyanos is called male and the other female, and the male is the darker of the two.*

The curious ancient distinction of sex in precious stones, which is mentioned in other early works, was apparently not connected with theories about the origin of stones, or even with the belief that certain stones had the power to generate others. A possible exception may be the supposed connection between the properties of the *lyngourion* and the sex of the animals alleged to produce the two varieties (sec. 28); Theophrastus is perhaps referring to this again in the present section. However, this connection should probably be regarded as accidental; for it was explained in the notes on section 28 that in all likelihood *lyngourion* was not an inorganic substance but either amber in general or a special variety of amber. It seems significant that similar modes of origin are not given for any true precious stone. It is likely that the concept of sex in stones was current long before the invention of the story about the *lyngourion* and was actually one of the sources from which it originated. In general, this concept was used to distinguish varieties of the same precious stone on the basis of their color, their relative brilliancy, or some other distinctive property. Thus, as he shows by his statements in this and the preceding section, Theophrastus relates sex in stones to transparency or color, assigning the male sex to the kind that are darker in color and the female sex to the ones that are paler. Pliny relates sex in stones both to color and to the relative brilliance of varieties of the same stone. For example, he describes the kinds of *carbunculus* in this way: *Praeterea in omni genere masculi appellantur acriores et feminae languidius refulgentes.*[217]

[217] XXXVII, 92.

(Moreover, in each kind the more brilliant are called male and those that shine with a fainter light are called female.) He also makes a similar statement about *sandastros*.[218] In discussing *sarda*, whose varieties Theophrastus distinguishes only by their color and transparency, Pliny has this to say: *et in his autem mares excitatius fulgent, feminae pigriores et crassius nitent*.[219] (Among these stones, too, the males glow more brightly, but the females are rather dull and shine with a dimmer light.) It will be seen, therefore, that according to this ancient concept of sex all stones of a certain type whose characteristic properties were more pronounced were called male, and those whose properties were less pronounced were called female. Though Theophrastus is apparently the earliest known writer to use this concept, his phraseology indicates that he did not invent it but merely incorporated it into his treatise as a method of distinction commonly understood in his day.

Theophrastus also varies the grammatical gender of the word used for stone, which sometimes appears as ὁ λίθος and sometimes as ἡ λίθος. Although the occurrence of this word in the treatise has been carefully studied, no clear distinction in meaning between the two genders has been found, and the problem is not solved by the suggestion in Liddell and Scott's lexicon that the feminine usually applies to some special stone. What is probably near the truth is that the grammatical gender of the word was unsettled in the early formative period of the language, so that the word was used in either gender interchangeably. Efforts were later made to straighten out the confusion by assigning special meanings to each gender. Galen shows that the argument not only concerned the word λίθος but extended also to ὁ πέτρος and ἡ πέτρα, the term used for hard stones and rocks; he points out the futility of the attempted distinctions and shows his impatience in the following words: ἔμπαλιν δὲ τὴν πέτραν λέγουσιν θηλυκῶς, οὐ τὸν πέτρον ἀρρενικῶς. . . . ἐγὼ γοῦν ἐξεπίτηδες εἴωθα, μεταβάλλων τὰ ὀνόματα, λέγειν ἑκατέρως ἅπαντα τὰ τοιαῦτα, περὶ ὧν ἀχρήστως ἐρίζουσιν ἔνιοι, δεικνὺς ἔργῳ μηδὲν βλαπτομένην τὴν σαφήνειαν τῆς ἑρμηνείας, ὁποτέρως ἄν τις

[218] XXXVII, 101.   [219] XXXVII, 106.

εἴπῃ.[220] (And again they speak of stone in the feminine, not of stone in the masculine. ... At any rate, my custom is purposely to interchange the names, and to adopt both methods of describing everything of that sort, which some people argue about so unprofitably; for I show that in fact the clearness of the interpretation is not harmed at all, whichever way one describes them.)

31. *kyanos*.

Theophrastus uses the word κύανος to designate two quite different types of material. In the present passage and in section 37, it evidently means a blue precious stone; but in sections 39, 40, 51, and especially 55, it is the name of certain natural or artificial substances that were used as blue pigments. There is also the stone called σάπφειρος, but the statements of Theophrastus in section 37 and the descriptions of Pliny[221] show that *kyanos* and *sappheiros* were simply varieties of the same mineral. Since the latter stone, as is explained in the notes on section 37, was almost certainly a variety of lapis lazuli, it is equally certain that the stone called *kyanos* was also a variety of lapis lazuli; the difference apparently was that *sappheiros* was the name given to the mineral when it contained numerous scattered specks of iron pyrites, whereas *kyanos* was the name used for the stone of solid blue color, or at least the stone in which iron pyrites were not present in noticeable proportion. Lapis lazuli ordinarily contains various other impurities such as mica, calcite, amphibole, and diopside, in addition to the intense blue mineral, lazurite, a complex sodium aluminum sulfosilicate, which determines the color of the mixture as a whole. In proportion to its content of lazurite, the color of lapis lazuli varies from a very deep blue to a light or even greenish blue. Probably Theophrastus is referring to these differences in depth of color when he distinguishes the varieties of *kyanos* according to sex. Though he lists *kyanos* among the stones used for seals, lapis lazuli, which is frequently intersected by hard crystals formed of minerals other than lazurite, is not very suitable as a material for engraving, and its scarcity in modern collections of ancient engraved gems shows that it was not often used for

---

[220] *De simplicium medicamentorum temperamentis ac facultatibus*, IX (Kühn ed., XII, 194).
[221] XXXVII, 119.

COMMENTARY

this purpose. However, it was extensively employed in ancient times for various ornamental purposes, as is shown by the numerous specimens that have survived. In Egypt, for example, it appears to have been used ever since predynastic times for beads, amulets, and other small objects,[222] and very early specimens from the Aegean region are known. But since there is no ancient or modern source of lapis lazuli in the vicinity of the Mediterranean, the stone must have been imported from a considerable distance. Persia is often given as the source, but geological investigations appear to show that it was, at the most, only a minor locality for lapis lazuli, since no important deposits or indications of ancient workings are known.[223] The only definitely established source of lapis lazuli for the ancient world is the very remote mine at Serri-sang in the upper Kokcha Valley between Parwara and Lower Robat, Badakshan, which some think was the only commercial source in ancient times.[224]

31. onychion.

Though the description of this stone seems to fit what is now called onyx, a banded chalcedony in which the layers, alternately white and dark in color, lie in planes one above the other, the word *onychion* has been used in the translation because there are definite indications on other grounds that ὀνύχιον had a broader meaning than the English word *onyx*. There seems to be no reason to doubt that the stone was banded chalcedony, since this is practically the only kind of striped stone to be seen among surviving examples of ancient engraved stones, but the term *onychion* might have included striped chalcedony in which the layers are not flat but angular, wavy, or concentric, varieties to which the name agate is now assigned. The descriptions of onyx given by Pliny,[225] though in some respects obscure and contradictory, definitely show that at least in his day the term not only included our onyx but striped agate also. That eye-agate, for example, was classified as a kind of *onyx* is very clear from his account. Pliny's

[222] Lucas, *Ancient Egyptian Materials and Industries*, p. 456.
[223] Partington, *Origins and Development of Applied Chemistry*, pp. 293, 416.
[224] K. Brückl, *Neues Jahrbuch für Mineralogie, Geologie und Paläontologie*, LXXII, Abt. A (1936), 37-56; R. J. Gettens, *Alumni (Revue du Cercle des Alumni des Fondations scientifiques à Bruxelles)*, XIX (1950), 342-57.
[225] XXXVII, 90-91.

use[226] of the word *onyx* for alabaster or onyx marble is another indication of the broad sense of the ancient term. Blümner[227] suggested that the word *onychion* might have had a more restricted meaning at the time of Theophrastus, or that Theophrastus was only acquainted with a particular form of banded chalcedony. This does not seem likely, however, since all the common varieties of striped chalcedony were in use as gems at the time of Theophrastus, and Theophrastus lists in his treatise nearly all the stones then used for the purpose. Since he does not distinguish onyx, sardonyx, and agate by special names, the best explanation is that *onychion* was a general term that applied to these banded chalcedonies as a class. So far as we know, Theophrastus is the first writer to mention and characterize such stones.

Banded chalcedonies of all sorts were very popular in antiquity as engraved stones, and some of the most beautiful ancient examples were executed in onyx or sardonyx. These were naturally the most suitable stones for large cameos, and a few remarkable specimens have come down to us.

31. *achates.*

Although the English word agate was derived from ἀχάτης and the Greek word has often been translated in this way, it seems reasonably certain that the ancient name did not have the same significance. Though he mentions its beautiful appearance, Theophrastus unfortunately gives no descriptive detail by which the stone can be identified, nor does Pliny in his account of *achates*[228] give any clear description of the stones included under this term. For the most part he merely lists varieties, though the names of these, being based upon color or some other distinctive property, afford some definite information. His brief characterization of *dendrachates*, for example, indicates that *achates* included at least one kind of true agate, the distinctive dendritic or moss agate, and this identification receives confirmation from the beautiful description of a stone with a similar name (ἀχάτης δενδρήεις) contained in the Orphic poem *Lithica*[229] of about the fourth cen-

---

[226] XXXVI, 59-61.
[227] *Technologie und Terminologie der Gewerbe und Künste bei Griechen und Römern*, Vol. III, pp. 266-67.
[228] XXXVII, 139-41.   [229] 232-38 (230-36).

COMMENTARY

tury of our era. It is certain, however, that all the varieties of our agate were not included under the term *achates*, for Pliny[230] clearly describes banded agate and eye-agate as varieties of *onyx*. Blümner,[231] who thinks the ancient term was restricted to certain kinds of agate, is probably not far from the truth, though it is by no means certain that only variegated agate was included under the term *achates*; for it seems evident from the descriptions of Pliny that some stones were then included that would not be classified as agates at the present time. Since the ancient systems of mineral classification were based upon appearance rather than composition, it is likely that any kind of attractive stone in which irregularly placed spots, streaks, veins, or other markings appeared against a background of contrasting color would have received the same general name, though probably it was usually applied to irregularly marked chalcedony, and possibly to jasper, for these forms of quartz are the most abundant and generally the most attractive of the variegated stones.

31. *the river Achates.*
According to the studies of Holm,[232] this river was probably either the modern Carabi or the Cannitello in southwestern Sicily.

32. *Lampsakos.*
This was a celebrated Greek settlement in Mysia, on the Hellespont.

32. *Astyra.*
The name of the town to which the stone was sent is uncertain. The manuscripts have στιράν or στιρράν, which cannot be correct. Turnebus changed this to Τίραν, and De Laet suggested that Τύραν would be better. But it is unlikely that Tyre, the city in Phoenicia, is meant, as that was on the coast and the participle ἀνενεχθείσης, which means "carried up," suggests that the stone was taken to some inland town. Actually, the Greek for Tyre is Τύρος; thus Τύρῳ appears in section 25. Furlanus proposed the

[230] XXXVII, 90-91.
[231] *Technologie und Terminologie der Gewerbe und Künste bei Griechen und Römern*, Vol. III, pp. 261-62.
[232] A. Holm, *Beiträge zur Berichtigung der Karte des alten Siciliens* (Lübeck, 1866), p. 15.

accusative of σφῦρα ("a hammer"). Schneider suggested Ἄστυρα, which is close to the manuscript reading. This form is the accusative neuter plural. Astyra was a town in the Troad, inland from Lampsakos, and there was another place of the same name near Antandros in Mysia.

Professor Gilbert Highet suggests that Στάτειραν should be read. This would mean that the stone was carved into a portrait of Stateira, who was the wife of Alexander the Great.[233] According to this interpretation, ἀνενεχθείσης would mean "when the stone had been brought up from the mine."

32. the King.

That the king was Alexander the Great is indicated by a passage in Pliny which reads as follows: *Gemmae nascuntur et repente novae ac sine nominibus, sicut olim in metallis aurariis Lampsaci unam inventam, quae propter pulchritudinem Alexandro regi missa sit, auctor est Theophrastus.*[234] (New gems which have no names are also produced unexpectedly; for example, Theophrastus reports that a stone was once found in the gold mines at Lampsacus which was sent to King Alexander on account of its beauty.) However, it is not certain that Theophrastus does mean Alexander; Pliny may have added the name on his own authority.

33. anthrakion.

Since ἀνθράκιον is evidently a diminutive of ἄνθραξ, Hill[235] and later commentators[236] thought that it might be the name of an inferior kind of garnet, but the hints that Theophrastus gives about the nature of the stone do not support this identification. In the notes on section 18 it was pointed out that the word *anthrax* meant originally a glowing live coal, and was applied later to stones of a similar red color; but since this word was also used to denote charcoal, it is equally reasonable to believe that the derived form *anthrakion* was one applied to very dark or black stones. The statements made by Theophrastus agree with this interpretation, for the *anthrakion* found at Orchomenos is described as darker than another kind called the Chian, and this

---

[233] Plutarch, *Vitae, Alexander*, LXX.   [234] XXXVII, 193.
[235] *Theophrastus's History of Stones*, pp. 88-89.
[236] E.g., Moore, *Ancient Mineralogy*, p. 208.

COMMENTARY

suggests that it was very dark in color, especially as the only Chian stone mentioned by ancient authors was apparently a black or dark rock variegated with spots or streaks of lighter-colored mineral matter. It will be recalled that Theophrastus alludes to this Chian stone in sections 6 and 7. What probably misled Hill and other commentators is that Pliny,[237] in quoting, or rather paraphrasing, this passage of Theophrastus, placed it at the end of his discussion of *carbunculus*, an error noted long ago by Beckmann[238] and by others before him. The use of the *anthrakion* of Orchomenos for making mirrors also shows that it was probably not garnet, for the latter, even when polished, is not distinguished by a high reflective power, nor is it ordinarily obtainable in pieces sufficiently large for such a purpose. Lenz[239] concluded that this *anthrakion* was obsidian, and though he gave no reason for his identification, it is possibly correct, for Pliny[240] speaks of a mirror of obsidian as though it were not uncommon and mentions immediately afterwards that obsidian was used by many for jewelry. When properly polished, this glassy rock yields a better reflective surface than any other dark stone, and the possibility that the ancients used obsidian for mirrors is supported by Beckmann, who remarks: "The image reflected from a box made of it, which I have in my possession, is like a shadow or silhouette; but with this difference, that one sees not only the contour, but also the whole figure distinctly, though the colors are darkened."[241] It is likely, however, that the obsidian mirrors of antiquity, being greatly inferior to those made of metal, were for ornament rather than utility. Against the identification advanced by Lenz it can be said that Theophrastus has already assigned another name, Liparean stone, to obsidian (sec. 14), but this may have been only a local name for the stone. Also, the *anthrakion* of Orchomenos may have been a darker variety of obsidian than Liparean stone, perhaps approaching pitchstone in density of color, and was possibly for this reason given a distinctive name. Of course, it is entirely possible that *anthrakion* was not obsidian at all, but some other dark or black stone that could be given a high polish.

[237] XXXVII, 97.
[238] *History of Inventions, Discoveries, and Origins*, Vol. II, pp. 67-68.
[239] *Mineralogie der alten Griechen und Römer*, p. 22.
[240] XXXVI, 196.   [241] *Op.cit.*, Vol. II, pp. 65-66.

Few attempts have been made to identify the stones from Troezen and Corinth which also came from the Peloponnesus. One suggestion is that they were varieties of *anthrakia* and therefore inferior kinds of garnet. Neither the general sense of the passage nor the remarks on the appearance of the stones support this interpretation, which may also be due to the careless quotation of Pliny. It seems far more likely that the stones from Troezen and Corinth were not *anthrakia* at all, but some other stones of attractive appearance. From the imperfect descriptions given by Theophrastus it is difficult to determine specifically what the stones were, but probably they were certain kinds of variegated quartz, such as colored jaspers, which are not uncommon in the general region around these two places.

34. *the remarkable ones are rare and come from a few places.*

Hill[242] in his translation assumes that Theophrastus is speaking here of places from which the more valuable kinds of garnets were obtained, but he seems to have misunderstood the real meaning of this section as well as the preceding one. Actually, Theophrastus is simply mentioning certain important foreign localities from which the more valuable kinds of precious stones were obtained.

34. *Syene near the city of Elephantine.*

Syene was situated on the east bank of the Nile just below the First Cataract. The site is now occupied by the town of Aswan. Elephantine was a city directly opposite on the southern end of the island of the same name. Though this district was known as the source of an inexhaustible supply of building stone, principally granite, there is no evidence that precious stones were ever found in the immediate vicinity of the two cities. Therefore, it would appear that Theophrastus is speaking of these places only as exporting points for precious stones that were found elsewhere in the country. Carthage and Massalia, mentioned just previously, are apparently referred to in the same way.

34. *Psepho.*

The name Ψεφώ, which is used here as a genitive, is not found

[242] *Theophrastus's History of Stones*, pp. 89, 91.

COMMENTARY

elsewhere. Salmasius suggested Ψεβώ as an emendation, and Hill adopted this but printed it incorrectly as Ψηβώ.[243] Psebo is a place mentioned by Strabo[244] in his description of Ethiopia as a large lake containing a populous island. Here the word appears in the nominative as Ψεβώ (Psebo). The modern name is Lake Tana, and the large island in it is called Dek Island. Possibly in ancient times this island or certain surrounding territory took the name of the lake, and it is to one or both of these that Theophrastus makes reference.

35. *Bactriana.*

An alternative translation is Bactria, the more usual name for this extensive ancient country. Though its boundaries are not entirely certain, it lay north of the Hindu Kush range, and most of it was between this mountain range and the Oxus River.

Previous commentators have generally assumed that these stones collected for mosaics were *smaragdoi*. Though it is true that *smaragdos* is mentioned in the first sentence of this section, they probably made this assumption because Pliny[245] says the *smaragdus* of Bactriana was found in essentially the same manner as the stones that are mentioned here. However, Pliny appears to have obtained his information from this very passage in Theophrastus, and in making his own version he introduced this mistake in identification. Pliny made a similar mistake when he supposed that the stones mentioned in section 33 of this treatise were varieties of the *carbunculus*. Clear evidence that he often misinterpreted the Greek of Theophrastus, or at least sometimes distorted his quotations, is contained in Book XXXVI, section 156, of the *Natural History*, where, after naming his authority, he gives a quite inaccurate version of a passage in the *History of Plants*. Bailey[246] discusses this particular case in detail. Apart from this statement of Pliny, which is found to be wrong, there is no reason for supposing that these stones found in the desert were *smaragdoi*.

There is nothing inherently improbable about the way in which the stones are said to have been found, since the sorting

[243] *Ibid.*, p. 90; C. Salmasius, *Plinianae Exercitationes* (Utrecht, 1689), 269 a G.
[244] XVII, 2, 3.   [245] XXXVII, 65.
[246] *The Elder Pliny's Chapters on Chemical Subjects*, Part II, p. 265.

· 133 ·

action of the wind in blowing away lighter sand particles would tend to leave behind the large grains and pebbles. It is probable that the fragmentary rock material collected in this manner consisted of different kinds of colored quartz.

36. *Among choice stones there is also the one called the pearl.*
The high value placed upon pearls by the ancients is evident from the statements of all early authors who touch on the subject. Pliny, for example, begins his elaborate account of the pearl[247] by saying that it holds the highest position among all objects of value, though in another place[248] he ranks it second among the precious stones.

The adjective διαφανής has been translated as "translucent," since "transparent," the usual meaning, is not appropriate to the pearl. Salmasius thought that the negative οὐ should be added so that the meaning would be "not transparent." Hill accepted this reading, but Wimmer did not. It does not seem necessary to change the text.

The eight Greek words in brackets (πλὴν . . . εὐμεγέθης) are not in the manuscripts or in Aldus but were taken by Schneider from the text of Athenaeus (III, 93), where they are attributed to Theophrastus.

36. *pinna.*
This name is now assigned to a genus of large bivalve mollusks which inhabit warm seas. The species common in the Mediterranean, to which Theophrastus is probably referring, was the first that was known, and so became the type for the genus. The Mediterranean pinna attains a length of about two feet. D'Arcy Thompson includes it in his list of Greek fishes under πίννη or πίνα (Latin *pinna* or *perna*), but he does not give an English name for it. He says that it is a pre-Hellenic word.[249]

36. *it is produced in India and certain islands in the Red Sea.*
Pearl fisheries are still operated today in the Red Sea, the Persian

---

[247] IX, 106.   [248] XXXVII, 62.
[249] *A Glossary of Greek Fishes* (St. Andrews University Publications, No. 45; Oxford University Press, 1947), p. 200.

COMMENTARY

Gulf, and along the coasts of India.²⁵⁰ In Herodotus the Red Sea ('Ερυθρὴ θάλασσα) meant the Indian Ocean, in which the Arabian Gulf ('Αράβιος κόλπος), now known as the Red Sea, was sometimes included. Later the term was also used for the Persian Gulf, e.g., in Diodorus (II, 11), and probably that is the meaning here.

37. *But there are some others.*
In this section Theophrastus mentions certain dull opaque ornamental stones. *Sappheiros* has already been listed twice (secs. 8 and 23) along with valuable stones, but the closing sentence of section 36 indicates that the other materials mentioned in this present section were rather common and not highly valued.

37. *fossil ivory which is variegated with white and dark markings.*
Though ὀρυκτός really means "dug up" rather than "fossil" in the usual modern sense, it would seem that both meanings are equally applicable in this case, for the material was apparently derived from the tusks, teeth, or, less probably, the bones of fossil animals. Possibly this even refers to ivory in the strict sense of the term, for the story of Pliny²⁵¹ about elephants burying their tusks seems to imply that the ancients recovered some ivory from the ground, though it is more likely that they dug up the tusks of mammoths and mastodons and that this was the real basis of Pliny's statement. Even in modern times a considerable amount of so-called ivory has been obtained from such fossil remains, particularly in Russia. The allusion to the mottled dark-and-white appearance of the material is readily understandable, since fossil tusks or teeth are often partly discolored by mineral or organic matter. Hill²⁵² pointed out that the word μέλανι in this passage does not necessarily mean "black," which is its most common meaning; for the same word is used immediately afterwards with reference to *sappheiros*, where the meaning is clearly "dark" in the sense of "dark blue." He therefore suggested that the meaning in this passage might also be "dark blue" and that the so-called

---

²⁵⁰ E. H. Kraus and E. F. Holden, *Gems and Gem Materials* (New York, 1925), p. 184.
²⁵¹ VIII, 7.   ²⁵² *Theophrastus's History of Stones*, pp. 94-95.

fossil ivory of Theophrastus was a substance of mottled blue-and-white color, not really ivory at all, but the mineral known as turquoise. Some later authors, e.g., Lenz[253] and Bailey,[254] have accepted this suggestion and have attempted to improve upon it by proposing that the material in question was not true turquoise, which is often light blue or green, but false turquoise or odontolite, which consists of fossil bones colored blue by vivianite, a hydrated iron phosphate. However, since all these identifications are based on the assumption that the material which Theophrastus calls fossil ivory was partly blue in color, they must be considered as little better than conjectures. As evidence against them it should be noted that Pliny discusses turquoise[255] and fossil ivory[256] in separate places, and when he mentions fossil ivory, he directly quotes this passage of Theophrastus and other passages about bones found in the earth and bony stones, most of which were clearly fossil remains. Though Theophrastus classifies fossil ivory as a stone, it does not follow that the identification that has been suggested is wrong, since elsewhere in the treatise he classifies amber and coral, also of organic origin, in the same way.

37. sappheiros.

It is certain that the σάπφειρος or *sapphirus* of the ancients was not the same stone as the transparent blue gem now called the sapphire. Theophrastus lists it here with other opaque minerals, and by comparing it with the *kyanos* implies that it was dark blue in color. Pliny, though apparently following Theophrastus in part, specifically states that it was blue and never transparent.[257] The statement of Theophrastus in section 23 that *sappheiros* seems to be spotted with gold is especially important for its identification. Pliny states in one passage[258] that eastern *sapphirus* was a stone in which gold sparkles, and in a second passage[259] that it was refulgent with spots like gold; and in a third passage[260] he compares it with another stone that was covered with drops of

[253] *Mineralogie der alten Griechen und Römer*, p. 23.
[254] *The Elder Pliny's Chapters on Chemical Subjects*, Part II, pp. 253-54.
[255] The *callaina* of XXXVII, 110-12.   [256] XXXVI, 134.
[257] XXXVII, 120.   [258] XXXIII, 68.
[259] XXXVII, 119.   [260] XXXVII, 139.

gold. Dionysius Periegetes[261] also confirms these descriptions, which can only apply to dark-blue lapis lazuli containing disseminated specks of iron pyrites. Owing to their appearance, these were naturally mistaken for gold by the ancients. This variety of the mineral is by no means rare. Lapis lazuli of solid color, or at least the kind that did not contain conspicuous amounts of pyrite, was given the name *kyanos*, as is explained in the notes on section 31. Pliny[262] remarks that *sapphirus* was not suitable for engraving when intersected with hard particles, and it is quite probable that because of the presence of pyrite, and possibly other hard crystalline minerals, the stone was rarely engraved but was simply used as a plain ornamental stone. Examples of ancient lapis lazuli speckled with iron pyrites have been found, and it is reported[263] that an ancient imitation made of blue glass containing grains of gold has even come to light.

37. *prasitis*.

The quantity of the middle vowel of πρασῖτις is uncertain. The Aldine text has πρασίτις, and this accent was accepted by Wimmer. In this same passage αἱματῖτις also appears as αἱματίτις. However, in the lexicon of Liddell-Scott-Jones both nouns appear with a long middle vowel like others of the same termination.

That *prasitis* was a green stone of some kind is clear both from its name and from the remark that Theophrastus makes about its color. The name was apparently derived from πράσον ("a leek"), and so πρασῖτις was probably a stone having a leek-green (dull dark-green) color. Theophrastus actually says that it was rusty (ἰώδης) in color, but this must refer, not to iron rust, but to the patina of bronze or copper rendered in the translation as "verdigris." In previous attempts to identify the stone, commentators generally have supposed that it was one of the various transparent or translucent green stones. Hill,[264] for example, thought it was "root of emerald," referring probably to prase; Lenz[265] identified

---

[261] 1105.
[262] XXXVII, 120.
[263] Partington, *Origins and Development of Applied Chemistry*, p. 118.
[264] *Theophrastus's History of Stones*, pp. 96-97.
[265] *Mineralogie der alten Griechen und Römer*, p. 23.

it as bluish-green fluorite; and Stephanides[266] suggested that it was beryl or chrysoberyl. That any of these identifications is correct seems very unlikely, for both the immediate context and the general content of the section show that Theophrastus is speaking of an opaque green stone. Though it is reasonably certain that *prasitis* was green and opaque, the lack of detailed description makes its specific identification almost impossible. Possibly it was the same as the *prasius* of Pliny,[267] who lists it as a common stone, one variety of which was evidently the same as our heliotrope, a kind of dark-green chalcedony or jasper marked with spots of red jasper. The statement about its color in the present passage tends to show that *prasitis* was green jasper, though the possibility still remains that it was serpentine or some other common green opaque stone. Like many other ancient mineral names, *prasitis* may have been a generic term and may therefore have denoted any dark-green opaque stone. Cylinders, seals, and various objects made of green jasper and other kinds of green opaque stone were in use in the Aegean region from early times, as is clear from the numerous finds that have been made.[268]

37. *haimatitis*.

All previous commentators have identified this as hematite, native ferric oxide; apparently they assumed that it was the same as the *haematites* of Pliny,[269] a term which certainly included most varieties of the mineral now called hematite. However, Pliny also mentions[270] a precious stone called *haematitis*, and it seems more logical to suppose that αἱματῖτις corresponded to this, for Theophrastus is clearly referring to a mineral substance of this kind. Furthermore, Theophrastus describes it in accordance with its name as having the color of dried blood, and Pliny also states that *haematitis* was blood-red in color; but these descriptions do not apply to the kinds of hematite that are hard enough for use as ornamental stones, since these varieties of the mineral are black, steel-gray, or, at the most, a dark-brownish red inclining to black. Only the streak that appears in these varieties and the soft compact

[266] *The Mineralogy of Theophrastus*, pp. 103-104.
[267] XXXVII, 113.
[268] Partington, *Origins and Development of Applied Chemistry*, pp. 325, 359.
[269] XXXVI, 144-47.   [270] XXXVII, 169.

COMMENTARY

kinds of hematite exhibit the pronounced red color from which its name was derived. Since Theophrastus is speaking here of an opaque stone, *haimatitis* could not have been any of the transparent or translucent red stones such as garnet or carnelian, though the practical certainty that these were always known by other names is itself sufficient warrant for rejecting them. The only common red opaque stone extensively used by the ancients was red jasper, an impure form of silica colored by ferric oxide, and it seems very probable that this was the *haimatitis* of Theophrastus. The allusion to its dull texture agrees well with the suggested identification, though it is probable that other stones of the same general appearance, such as red felsite, would have been given the same name by the ancients. Pliny[271] names Africa as a locality for *haematitis* and mentions its various uses as an amulet, all of which tends to support the identification here advanced, for red jasper is of common occurrence in northern Africa and it is known that the Egyptians obtained it locally and frequently used it for amulets and ornamental purposes.[272]

37. *xanthe*.

Though different in color, this stone was probably dull and opaque like the *haimatitis*, since the close relationship between the two stones implies that they were similar. Pliny[273] names *xuthos* as a stone of the same class as his *haematitis*, and it is almost certain that *xuthos* is the Latin equivalent, probably corrupt, of ξανθή, for Pliny seems on the whole to be following the statements that Theophrastus makes here. Since *haimatitis* was probably red jasper, it seems very probable that *xanthe* was yellow jasper. These two varieties of jasper are similar in all respects except color; even the color is due to iron oxide in both stones, but the difference is that in yellow jasper the oxide is in the hydrated form. Furthermore, red jasper and yellow jasper are often found together, sometimes in the same small piece, and this mode of occurrence probably explains why the close relationship between these two kinds of impure quartz was early recognized. Numerous ancient ob-

[271] XXXVII, 169.
[272] Lucas, *Ancient Egyptian Materials and Industries*, p. 454.
[273] XXXVII, 169.

jects of yellow jasper have been found, and since it appears to have been the only opaque yellow stone extensively used in antiquity, there is strong support for the identification given here. Yellow jasper was easily available to the peoples of the Mediterranean region. In particular, Lucas[274] mentions Sicily and the vicinity of Smyrna in Asia Minor as localities for this stone.

38. *coral*.

The spelling κουράλιον which appears in the text is a variant of the more usual κοράλλιον. In all probability, Theophrastus is referring here to the precious red coral, *Corallium nobile*, a species almost entirely confined to the Mediterranean. Both Dioscorides[275] and Pliny[276] clearly describe its characteristic form and color. It is this variety of coral which has been the most important commercially since very early times, owing to its striking color, luster, and fine texture. Red coral occurs frequently along the coasts and around the islands of the Mediterranean, and although the most important fisheries are now located off the African coast, it is still gathered in various areas along the northern side. The important localities are the coast of Provence, around Corsica and Sardinia, and in the vicinity of Naples and Genoa. In ancient times the northern localities were perhaps the only ones known, but Dioscorides mentions the promontory of Pachynos near Syracuse as the most important. This was the southernmost cape of Sicily, now known as Capo di Passaro. Pliny mentions various other localities, most of them around Italy. The manufacture of articles of red coral has for centuries been centered in Italy, particularly at Rome, Naples, and Genoa, and this is still true at the present day.

There is little literary evidence that the Greeks used coral to any extent for jewelry or other ornamental purposes. Since recent excavations of the sites of ancient Greek cities have yielded no coral ornaments,[277] it is likely that they did not value the material for such purposes. However, there is abundant evidence that

---
[274] *Ancient Egyptian Materials and Industries* (London, 1934), p. 347.
[275] V, 138 (Wellmann ed., V, 121).
[276] XXXII, 21-22.
[277] S. J. Hickson, *An Introduction to the Study of Recent Corals* (Manchester, 1924), p. 233.

COMMENTARY

it was highly prized as an ornament by various Oriental peoples in ancient times, and that it was extensively exported from the Mediterranean region as an important article of commerce. Coral appears to have been used by the Greeks and Romans mainly for medicinal purposes.[278] This belief in the special curative value of coral lasted well into modern times, for it was included in standard lists of drugs and in works on therapeutics as late as the last century.[279] Red coral has long been supposed to possess magical properties; it was therefore worn as an amulet in ancient times, especially by children,[280] and this practice has by no means disappeared today.

Apparently Theophrastus was not sure whether coral should be classified as a stone or as a plant. Pliny, on the other hand, was doubtful whether to classify it as a plant or as an animal; for though his descriptions lead one to suppose that he considered it to be a plant, his chapter on coral is included in his book on sea animals and the remedies derived from them. This illustrates the difficulty that naturalists have had until recent times in classifying coral. In his notes on this passage, Hill reflects the confusion that existed in his day when he says:

"The Nature and Origin of Coral has been as much contested as any one Point in natural Knowledge; the Moderns can neither agree with the Antients about it, nor with one another; And there are at this Time, among the Men of Eminence in these Studies, some who will have it to be of the vegetable, others of the mineral, and others of the animal Kingdom."[281] Hill's own conclusion, which he defends at length, was that coral is a plant, and he roundly criticizes those who think otherwise. But he changed his mind in his second edition, where he says it belongs to the animal kingdom. The animal nature of most of the corals was not understood until after the middle of the eighteenth century.[282]

[278] Dioscorides, V, 138 (Wellmann ed., V, 121); Pliny, XXXII, 24.
[279] E.g., A. Stillé and J. M. Maisch, *The National Dispensatory* (Philadelphia, 1880), pp. 464-65.
[280] Pliny, XXXII, 24.
[281] *Theophrastus's History of Stones*, pp. 97-98. But see 2nd ed., pp. 164-69.
[282] Hickson, *An Introduction to the Study of Recent Corals*, pp. 11-14.

38. *And in a way the petrified Indian reed is not very different in its nature from coral. But this is a subject for another inquiry.*

Indian reed itself (ἰνδικὸς κάλαμος), which is described by Theophrastus in his *History of Plants*,[283] appears to have been a species of bamboo. The meaning of the term "petrified Indian reed" is very uncertain. The allusion may be to bamboo or some other reed incrusted with calcareous sinter or to a true plant fossil. Among the works ascribed to Theophrastus by Diogenes Laertius[284] is a treatise *On Petrifactions* in two books. This lost work probably contained a systematic treatment of fossils as distinct from ordinary stones and minerals. The final sentence of this section may be an indication that this work was written after the treatise *On Stones*.

39. *Some of these contain gold and silver at the same time, but only the silver can be seen clearly.*

Such sulfide minerals as pyrite and galena often contain small amounts of gold and silver as impurities, and galena, which itself is silver-colored, is the chief modern source of silver. Since galena was also the source of the silver at the famous Laurion mines in Attica, it seems very likely that Theophrastus is alluding to it here. There is ample evidence that the citizens of Athens were familiar with the operations at these mines, and it is improbable that a philosopher like Theophrastus, with his special interest in scientific matters, would have had no technical information about the minerals and the processes used at the mines, especially since there was renewed activity in silver mining at Laurion[285] during the latter half of the fourth century, when this treatise was written.

The metallic appearance of certain natural sulfides is noted by both Dioscorides[286] and Pliny.[287] Though specific names are given by these authors to some natural sulfides that resemble metals in appearance, there is no evidence that a special name was given to any of these minerals at the time of Theophrastus.

[283] IV, 11, 13.
[284] V, 2, 42.
[285] T. A. Rickard, *Man and Metals* (New York, 1932), Vol. I, pp. 397-98.
[286] V, 115-17 (Wellmann ed., V, 100-102).
[287] XXXIV, 121.

COMMENTARY

39. *They are rather heavy in weight and have a strong odor.*

The sulfides are distinguished as minerals by their moderately high specific gravity. The mention of odor also indicates that sulfide minerals are meant. Though the pure minerals are odorless, on oxidation they easily give rise to sulfur dioxide, and this has a characteristic odor which is sharp and disagreeable. The strong odor that arose from heating sulfide minerals was certainly well known to the ancients, as the operation must have been frequently performed in mining districts. At Laurion, for example, it is very probable that galena was roasted on a large scale as a preliminary step in the reduction of the ore.[288] Furthermore, certain natural sulfides, such as marcasite, yield traces of sulfur dioxide and hydrogen sulfide when they are exposed to the weather, and their odors become strong when the minerals are broken. Both Dioscorides[289] and Pliny[290] allude to the disagreeable odor of certain partially decomposed sulfide minerals.

39. *There is also natural kyanos which contains chrysokolla.*

Theophrastus has just mentioned certain minerals or ores that resemble metals; he now proceeds in a characteristic way to mention certain ores that are unlike metals in appearance. From the notes on section 26 it will be seen that χρυσοκόλλα was a name used by Theophrastus for any bright-green copper mineral of an earthy nature. In this passage when he speaks of a blue mineral that contains a green copper mineral, the association of the two helps to fix the identity of both, since the only two ore minerals of these colors that are combined in this way are azurite, a dark-blue basic copper carbonate, and malachite, a bright-green basic copper carbonate. These two are very commonly found together, often intermingled or superimposed on each other. Forms are also known that contain an inner core of malachite surrounded by azurite. Beautiful specimens of intermingled azurite and malachite have been found at Laurion. Malachite is the more abundant mineral of the two and is often found alone; with or without the

[288] Rickard, *Man and Metals*, Vol. I, p. 381.
[289] V, 118 (Wellmann ed., V, 103). [290] XXXIV, 120.

associated azurite, it was almost certainly the most important source of copper in antiquity. It should be especially noted that Theophrastus, in speaking here of κύανος αὐτοφυής (native or natural *kyanos*), means a particular mineral. For information on the various natural and artificial products that were included under the general name *kyanos*, see the notes on section 55.

39. *There is another stone which is like glowing coals in color.*

The Greek is ambiguous here, since the word ἄνθραξι can refer either to charcoal, presumably in the form of glowing coals, or to a red precious stone. This red ore or mineral was probably cuprite, native cuprous oxide, which may be deep red or nearly black in color. So far as color is concerned, the allusion might be to pyrargyrite, otherwise called ruby silver or dark-red silver ore, a silver sulfantimonite. However, cuprite is more likely, because in the same passage Theophrastus mentions two other ores of copper, both of which are commonly found associated with cuprite.

40. *In general a great many unusual types of such stones are found in mines.*

Theophrastus evidently knew that a large number of mineral species existed, and the few that he mentions here and in later sections appear to be given mostly as examples of certain classes. The distinctions which he makes about the physical form or properties of mineral substances also appear to be given only as examples. His concluding remark in this section shows that he knew that other distinctions were possible.

41. *the Magnesian stone.*

It is evident from the descriptions given by Pliny and other ancient authors that various minerals of different chemical composition came from localities bearing the name Magnesia and were named after them. A white kind which Pliny[291] says was somewhat like pumice and which came from Magnesia in Asia Minor apparently corresponds to the one that Theophrastus mentions here. Various conjectures have been made about the identity of

[291] XXXVI, 128.

COMMENTARY

this stone. Blümner[292] thought it was magnetite, and Stephanides[293] suggested marcasite, but both identifications are improbable, for Theophrastus is speaking here of a stone that can be worked on the lathe, and the brittleness and hardness of both these minerals would prohibit the use of such a technique, especially with the appliances available to the ancients. Stillman[294] suggested that it might have been marble, dolomite, or gypsum. Though it is not improbable that gypsum in the form of alabaster might be intended, the most likely conjecture was made by Moore,[295] who believed that the Magnesian stone of Theophrastus was a form of talc. This identification best fits the conditions to be fulfilled: it was a soft mineral substance that could be easily turned on a lathe, it was available in large pieces, and it was white and in some way resembled silver. Some varieties of this hydrous magnesium silicate exhibit a characteristic white pearly or silvery luster. The stone mentioned in the next section was probably an impure form of the same mineral.

42. *In Siphnos there is a stone of this kind which is dug up about three furlongs from the sea.*

Siphnos is an island in the Aegean Sea northeast of Melos. Three stades, here translated as furlongs, are equivalent to about 1,820 feet. Pliny[296] is the only other ancient author who describes the stone found on Siphnos, and he appears to have obtained his information about it mainly from the present passage, though he adds that the green stone found at Comum (modern Como) in northern Italy was put to the same uses. It seems probable that this stone found on Siphnos was the variety of impure steatite or soapstone called potstone, usually greyish green to dark green in color and so soft that vessels of almost any shape can easily be carved from it. Though Fiedler[297] was unable to obtain any evidence that potstone occurred on the island, this does not prove it

[292] *Technologie und Terminologie der Gewerbe und Künste bei Griechen und Römern*, Vol. III, p. 278.
[293] *The Mineralogy of Theophrastus*, p. 159.
[294] J. M. Stillman, *The Story of Early Chemistry* (New York, 1924), p. 72.
[295] *Ancient Mineralogy*, p. 156.
[296] XXXVI, 159.
[297] Cited by Blümner, *Technologie und Terminologie der Gewerbe und Künste bei Griechen und Römern*, Vol. III, p. 66.

was never present there, since a fairly large local deposit could have been completely worked out by ancient operations.

42. *When it is heated in the fire and dipped in oil, it becomes very black and hard; and dishes for the table are made out of it.*

The blackening of the stone when it is fired and dipped in oil is a phenomenon that is readily understood. If the stone dishes were heated to the proper temperature and then dipped in a vegetable oil, which probably was the kind employed, black carbonaceous matter caused by the pyrolysis of the oil would be deposited on the surface of the stone and to some extent in the pores below the surface. A simple experiment was performed which showed how readily this takes place. One end of a green soapstone block ten centimeters long, three centimeters wide, and one centimeter thick was heated in a gas flame until it was red-hot, and then the entire block was plunged into olive oil. No change of color was observed during the heating, but when the block was plunged into the oil the heated part became black rapidly, while the relatively cool end of the block simply darkened slightly from absorption of oil by the stone. The blackening was more intense and occurred more rapidly in the place where the stone had reached the highest temperature. When the process had been repeated twice, the color of the part that had been treated became a deep black. Possibly this method of successive treatments was the one actually employed in ancient times, though Theophrastus does not say this. It was also noticed that the blackened stone had a somewhat greater surface hardness than either the untreated stone or the stone treated with oil at room temperature. The results of the experiment show that the statements of Theophrastus are quite accurate in this passage, as they are in other parts of the treatise whenever he seems to speak from his own knowledge. Probably the main purpose of treating the stone with oil was to harden it and make it less porous, though the process may also have improved its appearance. Dishes and other objects made of various kinds of steatite were widely used in ancient times, as is shown by the large numbers that have been found. Black examples are not uncommon, but apparently it is not certain whether their color

COMMENTARY

is due to the impure steatite from which they were made or whether it is the result of use or of some artificial treatment such as the one that has just been described.

43. *others can be carved with iron, but only with rather blunt tools.*

In spite of the lacunae in the text, it is easy to understand the paradoxes which Theophrastus mentions in this and most of the following section. He is evidently puzzled by the contrast between the hardness of mineral substances and the ease with which they can be split or broken. Many minerals or rocks cannot be scratched or cut by ordinary iron tools because of their surface hardness, yet because of their low cohesive strength they can be split or chipped easily by blows from a blunt iron tool. However, one might expect a priori that because they are hard they ought also to resist forces that tend to break them. As is well known, even diamond, the hardest of minerals, can be split with no great difficulty. When one mineral can easily scratch or cut another, the difference in their cohesive strength may sometimes be such that the harder one can be split more readily than the softer one, if the tool is made of a material with a lower surface hardness than either of the minerals. Theophrastus was apparently the first writer to call attention to these matters. It was not until the time of Mohs (1773-1839) and other modern mineralogists that differences in hardness became important criteria for classifying and identifying minerals.

The text in this section is so corrupt that it cannot be emended with certainty. But some interesting emendations suggested by Stephanides and others have been combined to form a possible reconstruction of it. These appear in a footnote referring to section 43 of the translation.

44. *the stone with which seals are carved.*

That this was corundum, native crystalline aluminum oxide, particularly the impure form called emery, is highly probable. The strongest argument for this view is that corundum was the only mineral available to the Greeks that was hard enough for engraving varieties of quartz or other hard stones that were commonly used for seals. Though some have maintained that diamond in the form of splinters or powder was used by the ancients for engraving stones, there is no real evidence to support this. It is very doubtful whether diamonds were even known to the peoples of the Mediterranean region in ancient times.[298] On the other hand, there is definite evidence, both literary and geological, that corundum in the form of emery was available to them and was employed for engraving gems. Dioscorides[299] mentions a stone named σμύρις as the kind used by jewelers for polishing their precious stones, and Hesychius[300] defines this as a kind of sand used for the same purpose. It is from this Greek word that our English word, *emery*, is derived. Pliny, however, appears to use *adamas* as one name for the kind of stone that was employed for engraving gems, for he states[301] that fragments of this were imbedded in iron by engravers and used for cutting the hardest substances; and in another place,[302] where he seems to be following in part the statements of Theophrastus in this section, he remarks that all precious stones may be cut and polished by the aid of *adamas*. Though his descriptions of *adamas*[303] show that the name was a generic one used for various minerals of pronounced hardness, it seems clear that corundum, either in the form of the well-crystallized mineral or in the impure form of emery, was usually meant. The *adamas* mentioned by Theophrastus in section 19 was probably this same mineral. Possibly Theophrastus does not give a definite name to the stone used for engraving seals because the corundum or emery that was used for this purpose went by so many different names. In another place, Pliny states[304] that the stone of Naxos had long been used for cutting and polishing precious stones, and though there is some confusion about the origin of its name, it is highly probable that it was described in this way because it first came from the island of Naxos in the Aegean Sea.[305] This makes its identification almost certain, because the chief mineral product of Naxos has long been a high grade of emery widely used as an abrasive. Later in the same pas-

[298] Partington, *Origins and Development of Applied Chemistry*, pp. 291, 507.
[299] V, 165 (Wellmann ed., V, 147). [300] *S.v.* σμιρίς.
[301] XXXVII, 60. [302] XXXVII, 200.
[303] XXXVII, 55-61. [304] XXXVI, 54.
[305] Blümner, *Technologie und Terminologie der Gewerbe und Künste bei Griechen und Römern*, Vol. III, pp. 198-99.

COMMENTARY

sage Pliny remarks that stones imported from Armenia afterwards replaced those of Naxos as cutting and polishing stones, but here Pliny has probably misinterpreted what Theophrastus says at the close of this section.

44. *whetstone*.
Though it is likely that the ancients generally used siliceous stones of various sorts for whetstones, it is possible that massive emery was also used for the purpose, for when Pliny[806] is discussing whetstones used with water, he says that the two best kinds came from Naxos and Armenia, and he then refers to his previous statements about their use for cutting precious stones. Certainly when Theophrastus speaks about the apparent identity of the whetstone and the stone used for engraving seals, he seems to indicate that massive emery was sometimes employed as a whetstone. However, since he had no reliable way of identifying two mineral substances of similar appearance or properties, his statement cannot be taken too seriously.

44. *And the ⟨best⟩ whetstone comes from Armenia.*
This statement is by no means free from difficulty. Schneider suggested the addition of ἀρίστη ("best") to the text, since the article ἡ cannot stand alone in this sentence. The sense requires αὕτη ("this stone") or ἡ ἀκόνη ("the whetstone"). The use of the feminine shows that Theophrastus is referring to αἱ ἀκόναι ("whetstones") in the previous sentence and not to ὁ λίθος ("the stone used for engraving seals"). Theophrastus probably did not mean to imply that Armenia was the source of the stone used for engraving seals; for neither corundum nor emery is known to occur as a commercial mineral within the boundaries of ancient Armenia. Though extensive commercial deposits of emery do occur in Asia Minor, these are all far to the west, mostly in the district around Smyrna.[807] It may perhaps be asked why a place as far away as Armenia is named as a source of the whetstone when suitable whetstones could certainly be obtained from places near the Mediterranean. Pliny[808] names several convenient sources like

[806] XXXVI, 164-65.
[807] Schmeiszer, *Zeitschrift für praktische Geologie*, XIV (1906), 188.
[808] XXXVI, 164.

Crete and Mt. Taygetus in Laconia as well as the distant sources like Armenia, Cilicia, and countries beyond the Alps. There is, however, nothing remarkable about whetstones being brought from a distant place like Armenia, since such stones may have been far superior in quality to those brought from nearer places. At the present day whetstones that have desirable properties and can be obtained only in certain places are often exported to distant markets.

45. *The nature of the stone which tests gold is remarkable, for it seems to have the same power as fire, which can test gold too.*
In this and the two following sections, Theophrastus alludes to the touchstone and its use in testing precious metals. Since these passages contain the earliest account of a method of determining the quantitative composition of an alloy, or, indeed, of a material of any sort, they are of considerable importance in the history of assaying in particular and of analytical chemistry in general.

The test of gold by fire is also mentioned by a few other ancient writers. Plato[309] apparently alludes to it, and Pliny[310] states that fire was used to test gold, and that the purity of the metal was confirmed if it retained its original color after being heated to redness. He adds that the term *obrussa* was used as the name for the test. In the *Leyden Papyrus X* the following description of the ancient test by fire is given in recipe 43:

A TEST FOR GOLD

If you wish ⟨to test⟩ the purity of gold, remelt it or heat it. If it is pure, it keeps the same color and remains pure like coinage after heating. If it appears whiter, it contains silver; if rougher and harder, it contains copper and tin; if black and soft, it contains lead.[311]

Pliny[312] gives details of the fire test as it was applied to silver, and in the *Leyden Papyrus X* the fire test for silver is described immediately after the one for gold that has just been quoted. It

---
[309] *Republic*, 413 E.   [310] XXXIII, 59.
[311] Translated from the text of Berthelot, *Archéologie et Histoire des Sciences*, pp. 284, 286.
[312] XXXIII, 127.

seems clear from these descriptions that the ancients applied this test in only a qualitative way. It could also have been used for rough quantitative measurements, since a similar procedure was once used in the French mints to estimate the composition of silver alloys,[313] and a somewhat similar method was used to estimate the percentage of copper in gold alloys in the old mints of Japan.[314] Possibly the ancients also used the fire test for rough assaying, though the available evidence certainly gives us no grounds for believing this. It is very unlikely, as some have conjectured, that in testing gold by fire they ever employed anything like the accurate modern method of fire assaying in which the gold is isolated by the fusion of a weighed sample of metal with chemical reagents, the silver is removed from the gold by acid treatment after this fusion, and finally the pure metal is carefully weighed.

45. the stone works by friction.
When gold or gold alloys are tested by being rubbed on a touchstone, a plainly visible streak of metal remains on the black surface of the stone. The intensity of the yellow color of this streak is directly related to the gold content of the metal.[315] In modern practice, streaks made by gold alloys of known composition are placed alongside a streak made by the metal that is being tested; this is done with touch needles, which are a graded series of heavy flat needles tipped with gold alloys. The color of the streak left by the metal is then compared with the colors of the standard streaks, and a match in color indicates that the metal has the same gold content as the corresponding standard alloy. In the modern use of the touchstone, the streaks left by the standard alloys and the unknown metal are nearly always tested by means of chemical reagents. With this refinement the method becomes more delicate and more reliable. There is no mention in ancient literature of the use of touch needles or of standard alloys in any form, but, unless standards of some sort had been used, the method would have produced only rough results. However, the statements of Theo-

[313] W. C. Roberts, *Journal of the Society of Arts*, XXXII (1884), 882.
[314] W. Gowland, *Journal of the Institute of Metals*, IV (1910), 11.
[315] No black streak is formed by gold alloys, even those of low fineness, as is erroneously stated by G. Thomson, *Classical Review*, LVIII (1944), 36. This misconception has been discussed by D. E. Eichholz, *Classical Review*, LIX (1945), 52.

phrastus in the next section and those made by Pliny[316] show clearly that the ancients achieved results in the use of the touchstone that approached in delicacy those obtained at the present time. Since the references made by ancient authors to the touchstone and its uses show that they had no direct knowledge of the method of performing the test, it is not surprising that no mention is made by them of any standards of comparison. It is very improbable that the ancients made this method of assaying more reliable by the use of chemical reagents, since the principal reagents that are necessary, namely, nitric acid and aqua regia, were almost certainly unknown to them. It is of considerable interest to note that the method of testing gold and gold alloys by the touchstone is still very widely used at the present time, especially by jewelers and by dealers in scrap gold.

Silver and silver alloys may also be tested by the touchstone, though the method is less suitable for silver than for gold. Theophrastus shows in the next section that the ancients did use the touchstone for testing silver. Although it was also used in early modern times for assaying the alloys of silver and copper, as is shown by the detailed descriptions of the process in various early works on assaying,[317] the method is no longer employed for the quantitative testing of silver alloys. It is still used, however, by jewelers and by dealers in precious metals to detect silver in various alloys, and appropriate reagents are applied to the streak left on the stone by the alloy of unknown composition. In the earlier use of the touchstone for assaying alloys of silver and copper, graded touch needles composed of alloys of silver and copper were employed, and the streak left on the stone by the unknown alloy was compared with those left by the standard alloys.

46. *They say that a much better stone has now been found than the one used before; for this not only detects purified gold, but also gold and silver that are alloyed with copper, and it shows how much is mixed in each stater.*

Probably the touchstone was first used only in a qualitative way

[316] XXXIII, 126.
[317] E.g., J. Pettus, *The Laws of Art and Nature in Knowing, Judging, Assaying, Fining, Refining, and Inlarging the Bodies of confin'd Metals* (London, 1683), pp. 63-65.

COMMENTARY

to distinguish fine gold and silver from their alloys, but when the technique of performing the test was improved and, as Theophrastus says here, a better kind of stone was discovered, this test was later developed into a method for determining the proportions of the precious metals in alloys. Though Theophrastus does not say so, the touchstone was undoubtedly used by the ancients to determine the proportions of gold and silver in their alloys. Pliny[818] discusses the different proportions of silver in native gold found in different places, and this information probably originated from assays made with the touchstone. Moreover, since early Greek coins made of gold, silver, or electrum show relationships of weight and value, it is clear that a method of assaying gold and silver alloys must have been known even before the time of Theophrastus. Probably the refined or purified gold mentioned in this passage should be understood to include not only fine gold but also the alloys of gold combined with silver, especially those of high gold content. The analyses that have been made of ancient gold objects show that even the best ancient gold contained a sensible proportion of silver, and much of it contained a considerable proportion. The invariable presence of silver in their gold seems to have been recognized by the ancients, for Pliny[819] remarks that all gold contains silver. The statement of Theophrastus may therefore be interpreted to include the use of the touchstone for assaying the alloys of gold combined with silver.

The importance of this passage as evidence that the stater was the real standard of reference in the Greek system of weights has been pointed out by Professor Ridgeway.[820]

46. *indications are obtained from the smallest possible weight. The smallest is the krithē, and after that there is the kollybos, and then the quarter-obol, or the half-obol; and from these weights the precise proportion is determined.*
The smallest proportion of metal that can be detected by the touchstone varies considerably and depends chiefly on the skill of the operator, the kind of alloy that is being tested, and the relative proportions of the metals in it. Though Theophrastus evidently

[818] XXXIII, 80.   [819] XXXIII, 80.
[820] W. Ridgeway, *Numismatic Chronicle*, Ser. 3, XV (1895), 104-109.

names in ascending order the amounts that can be detected in a sample of alloy weighing a stater, he does not give any precise information about the relative values of these weights. It is possible, however, to deduce both their relative values and their actual values in terms of modern weight units. The name of the second weight was also used for a small bronze coin, and the names of the third and fourth were used for two small silver coins, the quarter-obol and the half-obol. Since the stater, which is taken as the standard, was also the name of a silver coin, it seems reasonable to conclude that these weights were the same as those of the coins bearing the same names. The Attic silver obol weighed about 0.72 of a gram,[321] so that the largest weight was equal to about 0.36 of a gram, and the second largest to about 0.18 of a gram. The smallest weight, the *krithē*, literally the barleycorn, probably weighed about 0.06 of a gram, since in ancient systems of weights the smallest denomination was actually the average weight of certain seeds, and it can be shown by experiment that the barleycorn weighs on the average about 0.06 of a gram. The present English troy grain that weighs about 0.065 of a gram appears to have been originally based on the weight of the barleycorn.[322] The relative value of the second weight named by Theophrastus and its equivalent in modern units cannot be deduced with equal certainty. According to some authorities[323] on ancient weights and measures, the *kollybos* was perhaps equal to 1/32 of an obol, but this valuation gives it a weight of only about 0.02 of a gram, which, in terms of silver at least, is below the weight of the smallest denomination, the *krithē*. Since Theophrastus evidently lists the four weights in ascending order of magnitude, this weight seems inherently impossible for the *kollybos*, unless, indeed, Theophrastus is using the word *krithē* in a purely figurative sense to denote an exceedingly minute quantity. This seems very unlikely, however, since all the other denominations are clearly to be taken as real weights. Furthermore, it is highly improbable that Theophrastus would have mentioned a weight equal

---

[321] G. F. Hill, *Handbook of Greek and Roman Coins* (London, 1899), p. 64.
[322] Ridgeway, *Numismatic Chronicle*, Ser. 3, XV (1895), 104-109.
[323] F. Hultsch, *Griechische und römische Metrologie* (Berlin, 1882), p. 228; Hill, *Handbook of Greek and Roman Coins*, p. 64; B. V. Head, *Historia Numorum* (Oxford, 1911), p. 390.

COMMENTARY

to only 1/32 of an obol or 1/384 of a stater when he was discussing the touchstone method of assaying, because such a small difference in weight, if the stater were taken as the standard, would have been impossible to detect by this method. Actually, the evidence for placing the *kollybos* at 1/32 of an obol is outweighed by the contrary evidence furnished by this passage. As Ridgeway[324] has pointed out, the *kollybos* must have been a weight between the *krithē* and the *tetartemorion*, and probably had some simple relation to both these weights. Ridgeway concluded from a passage in Aristophanes,[325] from a definition of Hesychius,[326] and from numismatic considerations that the *kollybos* must have been equal to 1/8 of an obol, and it seems likely that this conclusion is correct. The relationship between the stater and these four weights, and the equivalents of these weights in grams, may therefore be tabulated as follows:

DENOMINATION	WEIGHT IN GRAMS	RATIO
Stater	8.72	1
Hemi-obolos	0.36	1/24
Tetartemorion	0.18	1/48
Kollybos	0.09	1/96
Krithē	0.06	1/144

It is very doubtful that the ancients, working without chemical reagents for enhancing the delicacy of the test, could have used the touchstone to detect a single grain of alloying metal in a stater of gold alloy, or 1 part in 144, as Theophrastus claims. It is even doubtful that they could have detected differences of 1 part in 144 in the composition of alloys that are less rich in gold, even though these are more easily assayed with the touchstone.

Professor Gowland,[327] who investigated traditional methods used by the goldsmiths of Japan, shows that when gold is alloyed with silver only and no reagent is used, a skillful assayer can obtain results with the touchstone that do not differ by more than 1 part in 100 from the results obtained by exact modern methods. Even this degree of accuracy can probably be obtained only on

[324] *Loc. cit.*
[325] *Pax*, 1200.
[326] *S.v.* κολλυβιστής.
[327] *Journal of the Institute of Metals*, IV (1910), 11.

gold-silver alloys that have a favorable range of composition, and it cannot be obtained on gold-copper alloys of any range of composition. Only under the most favorable conditions, including the use of reagents, is it possible to reach or exceed the degree of delicacy claimed by Theophrastus for the touchstone method of assaying. It is said to be possible under such conditions to estimate the gold content of alloys that are between 700 and 800 fine (70 to 80 per cent of gold) to within 1 part in 200.[328] The ancients were probably able to determine differences in composition represented by the other weights named by Theophrastus when the method was applied to gold alloys. It is especially interesting to note that, when the stater is taken as the standard, the two largest weights, the *tetartemorion* and the *hemi-obolos*, correspond to the half-carat and the carat, which are used on the present English and American commercial scale to express the quality of gold alloys. It is obvious, therefore, that by means of the weight scale given by Theophrastus, the ancient Greeks could have expressed the composition of their gold alloys in much the same way as it is done today.

For silver, the touchstone method of assaying is considerably less delicate than for gold, though Professor Gowland[329] shows that a skillful assayer can determine the proportion of the metals in a rich alloy of silver and copper to within 1.5 or 2 parts in 100. However, alloys poor in silver or very rich in it cannot be assayed with even this degree of accuracy. It is clear, therefore, that no difference in composition smaller than that represented by the *tetartemorion* could have been determined by the ancients when they were assaying silver alloys with the touchstone, and it is likely that the method was frequently not sufficiently delicate to measure difference in composition smaller than that represented by the *hemiobolos* when the stater was taken as the standard.

47. *All such stones are found in the river Tmolos.*
No river of this name is mentioned by ancient writers on geography. Strabo[330] applies the name to the mountain in Lydia lying between the river Hermos on the north and the Kaÿstros or

[328] T. K. Rose, *Metallurgy of Gold* (London, 1915), p. 554.
[329] *Journal of the Institute of Metals*, IV (1910), 13.
[330] XIII, 4, 5 and 12-13.

COMMENTARY

Caÿster on the south. He[331] names the Pactolos as a river rising on Mt. Tmolos and flowing into the Hermos. Hence it would appear that Theophrastus is mistaken about the name of the river in which touchstones were found. On the other hand, if the text is corrupt, he may originally have spoken of the river that rises on Mt. Tmolos. It is also possible that one of the other rivers rising on Mt. Tmolos may have been known in ancient times by the name of the mountain and that the text as we have it is correct.

Though Theophrastus in this section and the two preceding ones does not give any name to the stone used in testing precious metals and their alloys, it should be noted that in section 4 he said that a stone of this kind was called ἡ Λυδή, and it seems clear from the name of the river mentioned here as the sole source of such stones that the name Lydian stone is to be understood. Owing to the ambiguity of the text in section 4, it is possible, however, that the name "Heraclean stone," which really refers to the lodestone, was also applied to the touchstone. An earlier Greek name for the touchstone was βάσανος,[332] and this name may also have been current at the time of Theophrastus. At any rate, all three names, or derivations of them, were used by ancient authors later than Theophrastus. Sometimes later writers also refer to this stone without using any distinctive name. An example of this usage occurs in the thirty-eighth recipe of the *Leyden Papyrus X*. It is interesting to note that mineralogists now describe the kind of stone best suited for use as a touchstone by names that are based directly on the ancient Greek names. In English, for example, this kind of stone is called basanite or, alternatively, Lydian stone.[333]

47. *They are smooth in nature and like pebbles.*
It is probable that the touchstones described by Theophrastus were simply rounded pieces of alluvial slate, since Boz Dagh, the ancient Mt. Tmolos, is comprised largely of gneiss and slate,[334] and of these two rocks only the slate could have been used for touchstones. Though Theophrastus states that touchstones all came

---

[331] XII, 3, 27; XIII, 1, 23; XIII, 4, 5.
[332] Pindar, *Pythian Odes*, X, 67; Aristotle, *Historia Animalium*, VIII, 12, 597B; Aristotle, *De Coloribus*, III, 793B; Theognis, 417.
[333] Dana, *System of Mineralogy*, p. 189.
[334] Pauly-Wissowa, *Real-Encyclopädie* (2) VI, pp. 1627-28.

from this one locality, it is evident that there were other sources of such an abundant material as slate, and even of alluvial slate. The wider distribution of the stone used by the ancients for touchstones is, in fact, specifically reported by Pliny.[335] However, Theophrastus may be right in his statement about the source of supply in Greece during his lifetime, since slate, particularly the kind suitable for use as a touchstone, is by no means a common rock on the Greek mainland or on the islands of the Aegean. Black slate is a satisfactory material for touchstones, though modern touchstones are generally made from a velvety-black form of jasper, which is a much better material.

47. *The top part, which has faced the sun, differs from the lower surface in its testing power and tests better than the other. This is because the upper surface is drier, for moisture prevents it from picking out the metal.*

This difference in the moisture on the upper and lower surfaces of the stone can only refer to the condition of the slate when it was picked up from the shores of the river or the dry parts of the river bed. It is easy to understand why the clean upper surfaces of such stones were superior to the lower surfaces, since the latter were not only damp but also probably impregnated with fine particles of clay which lubricated the surfaces and reduced the desired abrasive action of the stone. These remarks of Theophrastus probably do not refer to the use of touchstones in mints or in shops where objects of precious metal were made, but rather to their use for testing the quality of metal as it was being mined. Thus Theophrastus may be speaking of the touchstone used for testing the metal that was mined in or near one of the rivers of Mt. Tmolos. There is historical evidence that Mt. Tmolos was an important source of gold during the reigns of certain kings of Lydia, particularly Alyattes and his son Croesus (560-546 B.C.), though not later.[336] It therefore seems likely that the information which Theophrastus is giving here may go back to these very mining operations. Moreover, it is not unlikely that the touchstone method of assaying may have come into use at the mines of Mt.

---

[335] XXXIII, 126. See the notes on sec. 4 for a quotation of the Latin text.
[336] Strabo, XIII, 4, 5.

COMMENTARY

Tmolos when the gold deposits were being worked, especially since the gold that was found there was probably in the form of electrum which varied greatly in quality. This variation is shown by the composition of coins made of electrum that were minted during the seventh and sixth centuries B.C. in Lydia or the neighboring countries.[337] Certainly some such method of assaying must have been necessary for the determination of the gold content of electrum and its value in relation to pure gold or pure silver.

48. *Of such kinds are the special qualities and powers found in stones. Earth has fewer of these, though they are more peculiar; for it is possible for earth to be melted and softened and then hardened again.*

The second part of the treatise, dealing with the various earths, begins here. In this context the verb τήκεσθαι really describes a softening caused by the action of water, and μαλάττεσθαι means a softening caused by the action of heat. Therefore τήκεσθαι does not refer to the melting of earths in the modern sense but rather to their disintegration by water. These two different ways of reducing solid bodies to an apparent or real state of fluidity are discussed at length by Aristotle,[338] who makes clear by examples the special meanings usually attached to these two words in the scientific writings of the Peripatetic school. Moreover, the distinction that Theophrastus makes between earths and stones seems to be based on the philosophical views expressed by Aristotle about liquefaction and solidification. In general, it is based on the Aristotelian doctrine mentioned in section 3, that opposite effects are produced by opposite causes. Since earths were supposed to be formed by the action of fire alone (sec. 3), they could be disintegrated readily by the action of water, because water, the moist-cold, was the direct opposite of fire, the dry-hot, and hence exercised the opposite effect on the same kind of matter.[339] Earths could be disintegrated by water, since they were supposed to contain pores large enough to admit water particles, which could easily enter them and cause disintegration.[340] On the other hand,

[337] J. Hammer, *Zeitschrift für Numismatik*, XXVI (1908), 17-51; W. Giesecke, *Antikes Geldwesen* (Leipzig, 1938), pp. 16-83.
[338] *Meteorologica*, IV, 6-9.   [339] *Op. cit.*, IV, 6, 383A.
[340] *Op. cit.*, IV, 9, 385B.

stones supposedly could be formed by the action of either fire or cold (sec. 3), and even by both, according to Aristotle.[341] If they were formed by cold, that is, by the departure of heat after the original matter had been highly heated, they could not be dissolved by the action of water, the cold-moist, since this could not dissolve what the cold solidified. On the other hand, heat could not dissolve them either, because heat was active in forming them. Moreover, certain stones, or stony materials such as pottery, which were supposedly formed by the direct action of fire, could not in general be softened or dissolved by fire. Neither could they be disintegrated by the action of water, because their pores, in contrast to those of earths, were so compacted in the process of formation that water could not enter and bring about their dissolution.[342]

After an earth had been mixed with water, the mixture could be readily softened by heat, since it was thought that any kind of matter that contained a considerable proportion of water was fusible (cf. sec. 10). But the continued action of heat on a mixture of earth and water could expel the water and so lead, first, to the thickening of the mixture and, finally, to its solidification, as for example in the formation of pottery.[343] By this process earths were transformed into what was stone in its essential nature, and this "stone" could not again be brought to the fluid state by the action of either fire or water.

Though it was believed that most stones could not be melted by fire, certain stones were in fact observed to be fusible (sec. 9). This phenomenon was not really an exception to the doctrine that opposite effects are produced by opposite causes, since such unusual stones contained a certain proportion of residual water which caused them to be fusible.[344]

The infinitive μαλάττεσθαι is an emendation made by Turnebus and accepted by Schneider; the manuscripts have ἀλλοιοῦσθαι ("to be altered"). This emendation is very plausible, since τήκεσθαι and μαλάττεσθαι are followed by τήκεται μὲν and μαλάττεται δὲ in the next sentence.

[341] *Op. cit.*, IV, 6.   [342] *Op. cit.*, IV, 6, 383B; 8, 385A.
[343] *Op. cit.*, IV, 6-7.   [344] *Op. cit.*, IV, 10, 388b-389a.

COMMENTARY

48. *It melts ⟨along with⟩ substances which are dug up and which can be liquefied, just as stone also does.*
Since all fusible substances, as well as metals which were less easily fusible, were thought to be composed wholly or in part of water, these could logically be as effective as water itself in causing an earth to melt. The melting of a stony material on being heated along with metals is mentioned in section 9 and discussed in the notes on that section.

48. *It is softened, and stones are made from it. These include the variegated ones and other composite stones . . . ; for all of these are made artificially when they are fired and softened.*
Though it seems necessary to translate λίθους as "stones," the context clearly indicates that artificial stony materials are meant. For the purpose of scientific classification Theophrastus evidently makes little or no distinction between natural stones and materials that are artificially produced. This passage clearly refers to the manufacture of ceramic products of some sort in which an earthy substance such as clay was first "softened" with water and afterwards baked or fired. There are not enough descriptive details in the passage to tell what particular kinds of ceramic products are meant.

49. *And if glass is also formed, as some say, from vitreous earth, this too is made by thickening.*
Though the reading of the text is only ἐκ τῆς ὑελίτιδος, the context indicates that an earth is to be understood. Strabo[845] specifically states that a vitreous earth (γῆ ὑαλῖτις) was used in Egypt in the manufacture of colored glasses. He also mentions the use of a special sand (ψάμμος ὑαλῖτις) in the manufacture of glass. Some commentators have concluded that Theophrastus meant the sand used in the manufacture of glass, but this does not seem probable. Throughout this part of the treatise he is discussing earths in particular. If he had specifically meant a sand, it is likely that he would have used the special word for it, as he does in sections 21, 40, and 58.

[845] XVI, 2, 25.

It is evident that Theophrastus had no first-hand knowledge of the way in which glass was made. Though glass objects of many kinds were well-known objects of commerce at the time, his ignorance is perhaps excusable, since there is apparently no literary[346] or archaeological evidence that glass was manufactured in ancient Greece.

Although the process of thickening (πάχυνσις) may seem to have no connection with the processes of softening or firing, in the Aristotelian sense of the word it was closely related to these other processes and was often a necessary consequence of them. According to Aristotle,[347] "thickening" was the compacting of dry matter by the removal of moisture, and this moisture could be removed by the agency of fire. Thus firing was a means of thickening. Moreover, the phenomenon of softening might occur in the course of thickening an earthy substance by the action of fire. Theophrastus apparently believed that glass was made by subjecting a particular earth to the action of fire, which compacted it to form glass. Here again, Theophrastus is clearly following the doctrines of Aristotle.

49. *The most unusual earth is the one mixed with copper; for in addition to melting and mixing, it also has the remarkable power of improving the beauty of the color.*

Since it is not certain what noun should be understood with the feminine article ἡ, this passage as a whole has been explained in very different ways. Hill[348] evidently believed that Theophrastus was still alluding to glass manufacture, and he therefore concluded that the article referred to ὕελος ("glass"), which is mentioned in the preceding sentence. Furthermore, Hill concluded from a suggestion of De Laet that the word χαλκῷ ("copper") was originally χάλικι ("gravel"), and he not only made this unwarranted alteration in the text but translated it incorrectly as "flints." Hence his translation of this passage is radically different from the one given here. Various scholars have accepted Hill's view that the passage refers to glass, but most of them have not

[346] Mary L. Trowbridge, "Philological Studies in Ancient Glass," doctoral dissertation (University of Illinois, 1922), p. 133.
[347] *Meteorologica*, IV, 6, 383a.
[348] *Theophrastus's History of Stones*, pp. 118-19.

COMMENTARY

followed him in his change of the text.³⁴⁹ Stillman, for example, criticizes Hill for changing the text and gives the following translation: "But most peculiar is that ⟨glass⟩ which is mixed with copper, for in addition to the melting and mixing, it has the additional property of causing a beautiful difference in color." Stillman considers this passage to be of historical importance as the first reference in literature to the coloring of glass with copper. His interpretation of the passage is plausible, since the intentional coloring of glass with copper was evidently a very common practice in ancient times, as is shown by the results of many modern analyses of ancient colored glasses. Though it would seem more likely that the copper was added in the form of an oxide or of certain copper minerals such as malachite, it is clear from a statement in Pliny that it actually was used in ancient glass manufacture. After remarking that the restless ingenuity of man was not long content to make glass from sand and soda alone, Pliny names various ingredients that were added and then says: *levibus autem aridisque lignis coquitur, addito cypro ac nitro, maxime Aegyptio. continuis fornacibus ut aes liquatur massaeque fiunt colore pingui nigricantes.*³⁵⁰ (A fire of light, dry wood is used, and copper and soda, especially the Egyptian kind, are added. The smelting is done, like that of copper, in a series of furnaces. Dark masses of a rich color are obtained.) Evidently the copper was oxidized rapidly enough in such a process to change it more or less completely into oxide which combined with the silica. Stillman's interpretation is therefore not only possible but very plausible.

It is unlikely, however, that the feminine article refers to glass, for the word ὕελος (another form of ὕαλος) is masculine in the preceding sentence. Since the word is usually feminine, it is possible that the masculine article ὁ is a textual error, but there is no evidence for this. The word ὑελῖτις ("vitreous earth"), which occurs in the same sentence, is feminine, but it is unlikely that the article ἡ refers to this. It is most probable that it refers to the word γῆ ("earth") in the sentence beginning with the words αἱ δὲ τῆς γῆς in section 48.

---

³⁴⁹ Cf. Lenz, *Mineralogie der alten Griechen und Römer*, p. 24; Blümner, *Technologie und Terminologie der Gewerbe und Künste bei Griechen und Römern*, Vol. IV, p. 390; Stillman, *The Story of Early Chemistry*, p. 21.
³⁵⁰ XXXVI, 193.

There is an analogous passage in the pseudo-Aristotelian work *De Mirabilibus Auscultationibus* which is not at all ambiguous and suggests that the present interpretation is correct. This passage may be translated as follows:

> They say that the copper of the Mossynoeci is very brilliant and light in color, though tin is not mixed with it, but a kind of earth which occurs there is smelted with it. They say that the discoverer of the mixture did not instruct anyone else, so that the copper objects formerly produced in these regions are superior, whereas those made subsequently are no longer so.[351]

In his Latin translation, which he attributed to Turnebus, Schneider recognized that the article ἡ referred to γῆ when he wrote: *Singularis est proprietas terrae quae miscetur aeri*.[352] Rossignol[353] showed by a French translation of this particular passage that he also recognized it to have this meaning, and this was accepted without question by Rickard.[354] Moreover, Mieleitner[355] gives the same interpretation in his German translation. On the other hand, Mély[356] follows the interpretation of Hill. It is probable that much of the confusion about the meaning of the passage arose because editors did not always recognize the place where Theophrastus begins a new topic. In the text as it has generally been printed, section 49 does not seem to begin at the right place, and in the present translation this has been corrected by starting a new paragraph within section 49.

Many scholars have concluded that the passage just quoted from the *De Mirabilibus Auscultationibus* refers to the manufacture of brass by the calamine process, and both Rossignol and Rickard assume without question that Theophrastus is also referring to the manufacture of brass. This is improbable, even though the theory seems plausible. The two passages apparently refer to the same process, and both appear to contain a circum-

---

[351] Sec. 62.

[352] The actual words in the original translation of Turnebus are: *Singularis est suaeque proprietatis quae miscetur aeri*. Here the word *terra* does not appear.

[353] *Les Métaux dans l'Antiquité*, p. 254 (footnote).

[354] *Man and Metals*, Vol. I, p. 157.

[355] K. Mieleitner, *Fortschritte der Mineralogie, Kristallographie und Petrologie*, VII (1922), 440.

[356] F. de Mély, *Les Lapidaires de l'Antiquité et du Moyen Âge* (Paris, 1902), Vol. III, fasc. 1, p. 8.

stantial, though very sketchy, account of the manufacture of brass by the ancient calamine process in which the copper was alloyed by melting the metal in the presence of the zinc ore, calamine, instead of directly with metallic zinc. The theory seems even more plausible because it is known from a variety of evidence that brass was manufactured by this process in Roman Imperial times. However, no specimens of brass dating from the time of Theophrastus or belonging to the two centuries after the date of his treatise have ever been found. It is true that a few early specimens of bronze containing zinc have been found; the earliest of these comes from Gezer in Palestine and dates from the second millennium B.C.,[357] but such bronze appears to be of accidental origin and, strictly speaking, is not brass at all. The earliest true brass is apparently the Roman alloy which first appeared about the middle of the first century B.C. and was widely used for at least the next two centuries. It is possible that no specimens of Greek brass have been found because there has not been enough archaeological exploration or chemical analysis of ancient Greek metal objects. At any rate, objective evidence that brass was manufactured at the time of Theophrastus does not exist. There may be no such evidence, because brass was made during Greek times only in some obscure locality or for only a brief period. This view seems to be justified by the passage in the *De Mirabilibus Auscultationibus*. Therefore it might be possible to argue that, in spite of the lack of objective evidence, this passage in the treatise *On Stones* alludes to the manufacture of brass; if so, it can be regarded as the earliest allusion of definite date to the manufacture of this alloy, since the passage in the pseudo-Aristotelian work cannot be dated with certainty and is probably later.

However, both this passage and the one in the *De Mirabilibus Auscultationibus* can be interpreted in other ways. The "earth" that was smelted with the copper in order to change the color of the metal may not have been calamine but some other earthy mineral. It may, for example, have been an arsenic mineral such as realgar or orpiment, or some arsenic preparation derived from one or the other of these minerals. Of some significance, perhaps, is the word λευκότατον, which is used in describing the appearance

[357] R. A. S. Macalister, *The Excavations of Gezer* (London, 1912), Vol. II, p. 265.

of the metal in the *De Mirabilibus Auscultationibus*. This word may easily be translated "very white" instead of "very pale" or "very light in color," and if the word is so understood, then the metal could not have been brass. However, it could have been a copper-arsenic alloy, since such alloys, even when they contain a relatively small proportion of arsenic, are white in color, not yellow like brass or bronze. Alloys of copper and arsenic were known in the Aegean region from very early times, as has been shown by chemical analyses.[358] Moreover, two recipes in the *Leyden Papyrus X* give methods of whitening copper by treating the metal with arsenic minerals or with products derived from such minerals. One of these recipes (No. 23) reads as follows:

WHITENING OF COPPER

To whiten copper, so that it can be mixed with silver bullion in equal parts, and not be recognized: take Cyprian copper, melt it, add one mina of decomposed realgar, two drachmas of ironlike realgar, and five drachmas of lamellose alum, and melt ⟨again⟩. In the second fusion four drachmas, or less, of Pontic wax are added, and the mixture is ignited and poured out.[359]

Probably the decomposed realgar mentioned in this recipe was, at least in part, arsenious oxide prepared by roasting the mineral, and the iron-like realgar may have been native arsenic. The wax served both to reduce arsenic compounds, so that the arsenic would alloy with the copper, and to prevent the oxidation of copper and arsenic during melting and casting. It may be of some significance that the wax is described as Pontic, since the account in the *De Mirabilibus Auscultationibus* refers to the manufacture of "whitened" copper by people living in one part of Pontus. This suggests that the account refers to the manufacture of a copper-arsenic alloy, not brass.

Since this passage in Theophrastus is similar to the one in the *De Mirabilibus Auscultationibus*, it might be argued that it also refers to the manufacture of a copper-arsenic alloy. However, the two passages are not so much alike that this conclusion is entirely logical. Theophrastus actually says nothing about whitening the

[358] E. R. Caley, *Hesperia*, Supplement VIII (1949), 60-63.
[359] Translated from the text of Berthelot, *Archéologie et Histoire des Sciences*, p. 278.

COMMENTARY

copper or even about making it paler in color. He merely states that the treatment with the earth improves the color of the metal. Therefore, it seems rather more likely that he may be referring to a refining process in which impure copper or bronze was melted with some earthy substance in order to purify the metal and give it a bright metallic appearance. The purification of metals by melting them and treating them with earthy material was a common practice in ancient times, as is shown by the many recipes in the *Leyden Papyrus X* where such purification processes are described. Frequently the earthy material was of a bituminous nature, and this served to reduce oxidized metal back to the metallic state. It seems rather significant that in the next sentence Theophrastus mentions another peculiar earth which was certainly a natural bituminous material.

On the whole it seems impossible to establish with certainty the exact nature of the process that Theophrastus mentions in this passage. He may be alluding to the manufacture of brass or to the manufacture of a copper-arsenic alloy, but it is much more probable that he is alluding to a mere refining process in which bronze or copper, possibly in the form of crude or scrap metal, was melted in the presence of some earthy substance, perhaps a bituminous earth, in order to obtain clean metal of improved appearance.

49. *And in Cilicia there is a kind of earth which becomes sticky when it is boiled, and vines are smeared with this instead of birdlime to protect them from woodworms.* Though Theophrastus does not give a name to the peculiar earth that was used for treating vines, later writers generally give a specific name to an earth that was probably the same as this or very similar to it. Dioscorides[360] calls it γῆ ἀμπελῖτις ("grapevine earth"), but mentions that some persons call it φαρμακῖτις (*pharmakitis*). Galen[361] calls it simply ἀμπελῖτις, and the Latin name *ampelitis*, given by Pliny,[362] is obviously a mere transliteration of this Greek name.

[360] V, 180 (Wellmann ed., V, 160).
[361] *De simplicium medicamentorum temperamentis ac facultatibus*, IX (Kühn ed., XII, 186).
[362] XXXV, 194.

Not only the descriptions of its properties given by ancient writers but also the localities that they mention as its source show beyond doubt that it was a natural bituminous material of some kind. The clearest description is that of Dioscorides, who says: "One should select the black kind, which resembles long pieces of pine charcoal, and is splintery and somewhat shiny, and which, furthermore, does not dissolve slowly when it is ground fine and some oil is poured over it." As Forbes[363] rightly pointed out, this is a description of a pure bitumen or asphaltite. In particular, it corresponds closely to what is now called glance pitch, a kind of bitumen which is known to occur in small deposits in the very localities where the ancient material is said to have been found. The Seleucian Syria named by Dioscorides as the source was next to Cilicia, the district which Theophrastus names, and it included the Pierian Seleucia, which, according to Strabo,[364] was the source named by Posidonius. Even though these localities are all in the same general region, the identity of the material cannot be fixed, since small deposits of various other bitumens also occur in the coastal districts at the extreme northwestern corner of the Mediterranean.

It may be objected that Theophrastus would not classify a compact material like glance pitch as an earth, especially since he has previously classified other compact bitumens as stones. Nevertheless, he might have regarded it as an earth in a special sense, for Galen[365] remarks that such a material was called an earth only because it could be resolved into a "mud" with water. There are positive indications, however, that the material used on vines at the time of Theophrastus was only a clay or sand impregnated with asphalt, which clearly could have been called an earth, and that in later times a pure bitumen came into use; this was still called an earth because it was used for the same purpose as the earlier material and was of the same general nature. When Theophrastus speaks of boiling the earth that was applied to vines, this is an indication that it was not glance pitch, for this bitumen has a high softening or melting point and a very high boiling

[363] *Bitumen and Petroleum in Antiquity*, p. 19.
[364] VII, 5, 8.
[365] *De simplicium medicamentorum temperamentis ac facultatibus*, IX (Kühn ed., XII, 186).

COMMENTARY

point. On the other hand, some of the impure bitumens soften or fuse at a low temperature and contain a good deal of enclosed water. When this water is heated at a relatively low temperature, it is expelled and makes the mixture look as if it were boiling. The apparent difference in the method of preparing the material at the time of Theophrastus and at later times also indicates that it was an impure bitumen. Posidonius[866] states that before application it was mixed with oil, which would be necessary if a viscous material were to be formed out of glance pitch. Dioscorides[867] also mentions its solubility in oil, though he does not specifically state that it was dissolved in oil before it was applied to the vines. Theophrastus, it will be noted, says nothing about adding oil. Possibly he was unaware that oil was mixed with it, or possibly when he refers to a boiling process, he means that the material was dissolved in hot oil. On the other hand, with some impure bitumens no addition of oil would have been necessary to obtain a preparation suitable for vines. Therefore, it is not unlikely that the earth here mentioned by Theophrastus was a bituminous clay or sand, not the pure bitumen which appears to have been used in later times.

Though the main reason for applying a bituminous preparation to the vines seems to have been to catch harmful insects on its sticky surface, it is also possible that it had an important and perhaps unrecognized virtue as a fungicide.

50. *It would also be possible to determine the differences that are naturally adapted for causing earth to turn into stone; for those that are due to locality, which cause different kinds of savors, have their own peculiar nature, like those which affect the savors of plants. But it would be best to list them according to their colors.*

In his *Causes of Plants*,[868] Theophrastus makes a comparison between the juices or flavors in plants and in earths. He seems to imply here that differences in environment, which certainly cause differences in plants, may also cause differences in the properties of earths, and that it might be possible to classify earths on such

[866] According to Strabo, VII, 5, 8. [867] V, 180 (Wellmann ed., V, 160).
[868] VI, 3.

· 169 ·

a basis. But he does not give such a classification, probably because he did not find it practicable. Instead, he suggests that it would be better to classify them according to color. This is another indication that the attitude of Theophrastus toward scientific problems tends on the whole to be more practical than that of his immediate predecessors, Plato and Aristotle, both of whom were more interested in the philosophical aspects of these problems than in their practical solution. Although Theophrastus suggests that earths should be classified according to color, he does not follow his own suggestion very closely in his discussion of the various kinds of earths. For example, the two red pigments, cinnabar and red ochre, are described in widely separated sections. Again, white lead is described in one section, and other white earthy substances are discussed much further on in the treatise. The truth is that Theophrastus does not seem to have found any systematic method of classifying what he calls earths, though some order is evident in his arrangement. Thus the two ochres, red ochre and yellow ochre, are treated together, probably because they are found together and are otherwise related. White lead and verdigris are mentioned one after the other, apparently because they were manufactured in a similar way. In reality, Theophrastus groups the earths according to similarities in their use, mode of occurrence, or method of manufacture, rather than according to their color.

The manuscripts do not contain any reference to location or environment; Wimmer substituted τῶν τόπων ("places") for τοὺς τούτων, following a conjecture by Schneider. And Turnebus expresses the same idea when he writes *locorum succos* in his Latin translation.

50. *Moreover, some seem to have been set on fire and burnt, such as realgar and orpiment and others of the same kind. To put it plainly, all of these result from a dry and smoky exhalation.*

As was explained in the notes on section 3, Theophrastus closely follows the ideas of Aristotle about the origin of earths,[869] though

---

[869] For further discussion of the similarity of the ideas of Aristotle and Theophrastus on the origin of mineral substances generally, and for remarks on the interpretation of this particular passage, cf. D. E. Eichholz, *Classical Quarterly*, XLIII (1949), 143-46.

COMMENTARY

he is more specific about the origin of earths through fire. The concluding statements of sections 54 and 69 are of particular interest, since they express his belief that fire was the causative agent in the formation of certain earths. The final statement of Theophrastus in this present section is of special significance in the history of theories of combustion, since it is really a root idea underlying most of the theories promulgated before the time of Lavoisier. The famous and erroneous phlogiston theory of Becher and Stahl, for example, which was so widely accepted in the eighteenth century, embodied the same idea as a fundamental principle.

51. orpiment, realgar.

It is important to note that these mineral substances are nearly always mentioned together by ancient writers, since this helps to identify them as the two native sulfides of arsenic. Theophrastus has previously mentioned them together in section 40 and again in section 50. Both Dioscorides[370] and Pliny[371] mention them in successive sections. The names *arrhenicum* and *sandaraca* that Pliny uses are obviously mere transliterations of the Greek names used by Theophrastus. Orpiment, native yellow arsenic sulfide (As_2S_3), and realgar, native orange-red arsenic sulfide (As_4S_4), nearly always occur together and are often found intermingled in the same small specimen. Pliny alludes to the mixed form of the two minerals, and Dioscorides states that they are found in the same mines. This peculiar mode of occurrence is sufficient to explain why they are always discussed together by ancient authors. Since the *Leyden Papyrus X*[372] and other early technical works mention the use of one or the other of these minerals for whitening copper, it is almost certain that they were arsenic compounds, and the various descriptions of their physiological effects supply further evidence. The descriptions of their physical appearance given by Dioscorides and by Pliny are adequate to identify them as these particular arsenic minerals.

A few specimens of ancient orpiment and realgar have been found and positively identified as such by chemical tests. During the excavations at Corinth in 1926 Shear found a pottery vessel

[370] V, 120, 121 (Wellmann ed., V, 104, 105).   [371] XXXIV, 177, 178.
[372] Recipe No. 23. See the notes on sec. 49.

containing an orange-red pigment which was subsequently examined by Foster[373] and identified as realgar. A sample of a yellow pigment found in a Greek grave dating from the fifth or fourth century B.C. was identified by Rhousopoulos[374] as pure orpiment. It is likely that both orpiment and realgar were used as pigments long before the time of Theophrastus, although at present the evidence applies only to orpiment. This natural pigment has been found on Egyptian mural paintings and on various Egyptian objects dating from as early as the Eighteenth Dynasty, and a linen bag containing a small quantity of the mineral was discovered in the tomb of Tutankhamen.[375] Since orpiment and realgar occur much less frequently than other pigments on Greek objects found in excavations, it is probable that they were not very much used as colors in Greek times.

Though they occur at various places in Europe, it seems almost certain from the statements of ancient authors that the only ancient sources of supply were in Asia Minor or even farther to the east. Strabo[376] speaks of a realgar mine at Pompeiopolis in Paphlagonia, and Vitruvius[377] mentions that orpiment was dug up in Pontus. Dioscorides[378] also gives Pontus as a source of orpiment and mentions Mysia and Cappadocia in addition. Modern geological exploration has shown that orpiment and realgar occur at various places in Asia Minor.

51. red ochre.

Though there can be no doubt that the word μίλτος usually designated what we now call red ochre, a mixture of red ferric oxide with clay, sand, and other impurities, this was probably a general term, like so many of the other Greek names for minerals, which included all the pigments that owe their color to the presence of red ferric oxide. That it was applied to an artificial red iron oxide pigment is clear from what Theophrastus says in sections 53 and 54. His remarks in section 53 about a light-colored

---

[373] W. Foster, *Journal of Chemical Education*, X (1933), 276.

[374] P. Diergart, *Beiträge aus der Geschichte der Chemie dem Gedächtnis von Georg W. A. Kahlbaum* (Leipzig and Vienna, 1909), p. 187.

[375] Lucas, *Ancient Egyptian Materials and Industries*, p. 400.

[376] XII, 3, 40. The word in the text is *Pompeioupolis*.

[377] VII, 7, 5.

[378] V, 120 (Wellmann ed., V, 104).

COMMENTARY

variety, and the results of analyses of Greek iron oxide pigments, suggest that it was also applied to pale red or even pink pigments consisting of little more than clay or chalk colored with ferric oxide. On examining eight such pigments that were found on objects excavated at Athens, Midgley[379] discovered that six of them contained less than twenty per cent of ferric oxide, and that one of them, a pink pigment, contained only about three and a half per cent. Since no special name appears to have been applied to red iron oxide pigments artificially produced by roasting yellow ochre, or to the pale red or pink pigments, it is very likely that they went by the same name as red ochre itself.

Red ochre and other red iron oxide pigments occur so frequently on ancient Greek objects of all sorts, and so many vessels containing the remains of these pigments have been found, that they were probably used in Greek times more extensively than any other kind of pigment. Probably their great abundance and low cost were the principal reasons for their widespread use.

51. yellow ochre.

It seems reasonably certain that the word ὤχρα invariably designated what today is called yellow ochre, a mixture of hydrated ferric oxide with clay, sand, and other impurities. The information in sections 53 and 54 about the conversion of ὤχρα into μίλτος by roasting is almost decisive for this identification. Moreover, the chemical examination of ancient pigments of yellow ochre that were found in the excavations of the Agora at Athens has shown that they have the same composition as the mineral now called yellow ochre.

Theophrastus discusses the occurrence and the sources of red ochre, but he says very little about the sources of the yellow ochre used in his day. Probably this is because the principal source of the best yellow ochre was so well known to him and to his contemporaries that he considered it quite unnecessary to mention it. Most of the later writers on ancient pigments mention Attica as the source of the best yellow ochre. Dioscorides[380] actually implies that it was the only satisfactory source. Pliny[381] names other

---

[379] S. W. Midgley, "Chemical Analysis of Ancient Athenian Pigments," Senior thesis, Princeton University, 1936, pp. 14-21.
[380] V, 108 (Wellmann ed., V, 93).   [381] XXXIII, 158-59.

sources of yellow ochre, to which he gives the name *sil*, but states that the Attic mineral was the best. On the other hand, though Vitruvius[382] also names Attic ochre as the best, he states that it was not obtainable, and implies that this was because the Laurion mines were no longer in operation. Some yellow as well as red ochre can still be found at Laurion in the ancient mining district, and very good specimens have been collected in the modern workings. There are indications of old workings of iron minerals in various parts of Attica.[383] Yellow ochre also occurs elsewhere in Greece and on several of the Aegean islands. Pliny shows that it was mined in Greece outside Attica in ancient times; he names the island of Scyros as well as the province of Achaia as sources of a dark variety of yellow ochre.

Though modern excavation has shown that yellow ochre was less commonly used by the ancient Greeks than red ochre, it is clear that it was more extensively used than most of the other kinds of pigments. It was certainly the only yellow pigment that was in common use.

51. *chrysokolla*.

As was explained in the notes on section 26, this was a general name denoting any bright-green copper mineral that occurred as an earthy incrustation. It is clear from its inclusion here among the coloring materials that the ancient *chrysokolla* was used as a pigment and not solely as a "gold-glue," as its name suggests. There is definite archaeological evidence that the natural green copper carbonate known as malachite was used by the Greeks as a pigment. For example, Midgley[384] demonstrated that malachite was the green coloring material on a terra cotta object of the fourth century B.C. found in the excavations at Athens. It is likely that natural copper silicate, the modern chrysocolla, was also used as a green pigment, although this mineral has not actually been found among the Greek pigments that have been chemically investigated. On the whole, the rather marked scarcity of green copper pigments on ancient Greek objects tends to show that they were not much used as coloring materials by the Greeks.

[382] VII, 7, 1.
[383] Davies, *Roman Mines in Europe*, p. 252.
[384] "Chemical Analysis of Ancient Athenian Pigments," p. 25.

COMMENTARY

51. *kyanos.*
From the context this would appear to be the *natural kyanos,* blue copper carbonate, which Theophrastus has mentioned before in section 39. It was rarely used as a coloring material, so far as this can be decided by the examination of pigments found on ancient Greek objects. Thus the archaeological evidence confirms the remark of Theophrastus about the scarcity of the pigment. An artificial calcium copper silicate was much more widely used as a blue pigment by the Greeks. Detailed information about the blue pigments included under the name *kyanos* is given in the notes on section 55.

51. *yellow ochre can take the place of orpiment, since there is no real difference in their color, though there seems to be.*
Though both are yellow pigments, orpiment is actually more brilliant in tone than yellow ochre. Since no Greek portraits of the time of Theophrastus have survived, it cannot now be determined to what extent yellow ochre was used instead of the more brilliant orpiment. Probably it was used to a much greater extent, if one may judge from the pigments found on Egyptian and Roman mural paintings. Pliny[385] names Polygnotus and Micon as the first artists to use yellow ochre, and adds that they used only Attic ochre, though their successors used other kinds as well.

52. *But in some places there are mines that even contain both red ochre and yellow ochre together, as for example in Cappadocia.*
Red ochre and yellow ochre often occur together as well as in separate deposits. Modern geological surveys show that iron oxide minerals of various sorts are still to be found in considerable quantities within the limits of ancient Cappadocia in central Asia Minor.[386]

[385] XXXIII, 160.
[386] Schmeiszer, *Zeitschrift für praktische Geologie,* XIV (1906), 190.

52. *But they say that the risk of suffocation is a serious matter for the miners, since this can happen to them quickly and takes a very short time.*

Accidents were probably the result of inadequate timbering or other means of support in the galleries or shafts cut in soft, dangerous ground. That the roofs of ancient mines were often improperly supported is clear from ancient allusions to mining accidents and from modern explorations of ancient mines. Statius[387] likens the burial of a miner under a falling roof in the silver mines of Spain to the sudden crushing of a wrestler by his opponent. Gowland[388] reports that the skeletons of more than fifty men were discovered in some very old workings near ancient Iconium in Asia Minor; they had evidently been entombed in an underground chamber by the sudden collapse of the gallery leading to the open air. In section 63, Theophrastus describes how a white earthy mineral was mined on the island of Samos and shows clearly how dangerous it was to work in mines exploited for soft minerals. Though it is barely possible that Theophrastus may be alluding here to accidents caused by the presence of noxious gases, this is very unlikely, since such gases would normally be absent from ochre mines.

52. *The best red ochre seems to be that of Ceos.*

Ceos (Κέως), an island in the Aegean situated about fourteen miles southeast of the southern tip of Attica, is not mentioned by Dioscorides, Pliny, or Vitruvius as a source of the pigment. This suggests that the ochre mines on Ceos, though at one time yielding an excellent product, were exhausted before the beginning of the Christian Era. According to Davies,[389] it is still possible to see signs of the ancient mining of iron minerals at various places on the island. At Spathi, for example, there are said to be ancient stopes and galleries. At Oriko, three miles south of Spathi, many ancient stopes and adits, some containing ancient tools, were uncovered when the mines were reopened in modern times. Traces of ancient galleries only two to three feet high are said to be in

[387] *Thebaid*, VI, 880-85.
[388] W. Gowland, *Archaeologia*, LXIX (1917-1918), 157.
[389] *Roman Mines in Europe*, p. 257.

COMMENTARY

evidence at this place. The presence of black-glaze sherds near the site affords a clue to the date of the workings. Caves in the brown ironstone on the northeast coast may, however, have been the site of the principal ancient workings. Some interesting evidence concerning the ancient mining of red ochre on Ceos is given by an inscription belonging to the fourth century B.C. which records an agreement between Athens and Ceos for regulating the export of red ochre from the island; the terms provided that this could not be exported except to Athens and could be sent only on an authorized vessel.[390]

52. iron mines also contain red ochre.

This shows clearly enough that the term μίλτος was applied to earthy hematite containing a high proportion of ferric oxide as well as to red ochre in the modern sense. Midgley[391] found 48 per cent of ferric oxide in a deep red pigment taken from a terra cotta object discovered in the excavations at Athens. This is a high enough proportion to class the pigment as an iron ore. A brownish-red pigment found in the bottom of a broken vessel discovered in the Athenian Agora has been identified as earthy hematite.

52. the Lemnian kind.

Lemnos, a fairly large island in the northern Aegean Sea, midway between Mt. Athos and the Hellespont, is still known locally as the source of a particular kind of medicinal earth. The accounts of Pliny and Dioscorides suggest that the red ochre of Lemnos was the same as the famous medicinal earth of that island, though it is very doubtful that this is true. Pliny[392] mentions Lemnian *rubrica* as a pigment that was regarded by some authorities as the best of the red ochres, inferior only to cinnabar. But he goes on to say that every piece sold was officially sealed, and for this reason the name *sphragis* was given to it. These statements seem to identify this variety of red ochre as the Lemnian medicinal earth, since this is known to have been prepared in the form of tablets

[390] *Inscriptiones Graecae*, Vol. II, Part 1, No. 546. See also M.N.Tod, *A Selection of Greek Historical Inscriptions* (Oxford, 1948), Vol. II, No. 162, pp. 181 ff.
[391] "Chemical Analysis of Ancient Athenian Pigments," p. 14.
[392] XXXV, 33.

impressed with a seal. Dioscorides[393] discusses the Lemnian medicinal earth immediately after his paragraphs on Sinopic ochre and artificial red ochre but does not mention it in the part of his work devoted to other medicinal earths. That Pliny confused two quite different earths found on Lemnos, or at least two different kinds of red ochre, seems clear from the critical and detailed statements of Galen, who actually visited Lemnos to investigate the manufacture of the medicinal earth for which the island was then famous. According to Galen there were, in fact, three different earths found on Lemnos. One was the medicinal earth, another was a true red ochre suitable for use in painting, and still another was an earth used for cleaning clothes. Galen further remarks that some people called the medicinal earth μίλτος because of its color, though this was an incorrect designation. He clearly differentiates between the medicinal earth of Lemnos and the red ochre of the island when he says: "Though it has the same color as red ochre, it differs from it in not staining when it is touched...."[394] It may be concluded, therefore, that the Lemnian *miltos* mentioned by Theophrastus was a true red ochre suitable for use as a paint pigment, whereas the Lemnian medicinal earth, which he does not mention anywhere in this treatise, was probably a clay stained red with ferric oxide. This was used extensively as a popular remedy in various European countries until comparatively recent times.

52. *the one called Sinopic; this is really Cappadocian red ochre, but it is brought down to Sinope.*
Since Sinope, the modern Sinub or Sinop, had the only good natural harbor along the entire south coast of the Euxine, it was the principal port for the export of products from the whole eastern part of Asia Minor. None of these products seems to have been as widely known as the valuable red ochre named after the city. This was so famous that the term *sinopis* finally became a synonym for red ochre itself. Pliny[395] remarks that *sinopis* derived its name from the Pontic city of Sinope, and adds that it

---
[393] V, 113 (Wellmann ed., V, 97).
[394] *De simplicium medicamentorum temperamentis ac facultatibus,* IX (Kühn ed., XII, 169-70); *De antidotis,* I (Kühn ed., XIV, 80).
[395] XXXV, 31.

COMMENTARY

also occurred in various other places such as Egypt and the Balearic Islands. It has been suggested from time to time by various scholars that the *miltos* of Sinope was not red ochre at all, but the rarer and more costly red pigment, cinnabar. This opinion is maintained, for example, by Leaf,[396] mainly because it would have been so expensive to transport such a common product as red ochre through the difficult country lying between Cappadocia and Sinope that it could not have been sold at a profit in Greece, where it would have had to compete with the red ochre found abundantly in much nearer localities such as Ceos and Lemnos. Though some of his other arguments are also ingenious, it is not at all likely that his identification is correct. A very serious objection is that cinnabar does not occur within the confines of ancient Cappadocia, though various iron minerals such as brown iron ore and the ochres are found in many places.[397] Moreover, cinnabar is not found in the localities listed by Pliny as sources of *sinopis*, but red ochre occurs in these places. The argument that red ochre imported from Sinope would have been too costly to compete with the product from neighboring places is easily refuted. In the first place, natural red ochres differ greatly in quality and in suitability for use as paint pigments, so that an imported ochre of high quality could easily have been sold at a considerably higher price than ordinary red ochre. Even today, natural red ochres which have a particularly desirable hue, brilliancy of tone, or tinting strength are brought great distances to be marketed in direct competition with domestic ochres selling at a much lower price. Thus, for example, the red ochre found at Ormuz on the Persian Gulf is exported in large quantities to England, the United States, and other distant countries. Moreover, the Sinopic red ochre may have been a pigment of very high iron oxide content, so that much less was needed when it was diluted with white pigments to produce light-red or pink colors, and thus it may have been cheaper for this purpose than ordinary ochres selling for a third or a fourth as much. The remark of Dioscorides[398] that Sinopic *miltos* was liver-colored definitely suggests that

[396] W. Leaf, *Journal of Hellenic Studies*, XXXVI (1916), 10-15.
[397] Schmeiszer, *Zeitschrift für praktische Geologie*, XIV (1906), 190.
[398] V, 111 (Wellmann ed., V, 96).

it was just such an earthy hematite. Most of the scholars who would identify Sinopic *miltos* or *sinopis* with cinnabar either ignore the descriptions of these minerals left to us by ancient writers or misinterpret them. The descriptions of Theophrastus, Vitruvius, Pliny, and Dioscorides indicate clearly enough that the pigment exported from Sinope was not cinnabar.

53. *It is dug up by itself in* . . . .
It is difficult to translate ἐν τῷ μικρῷ ("in the small . . . "), for it seems to require some noun like "district" or "mine." It is unlikely that it can mean "in small quantities." Theophrastus uses κατὰ μικρά in this sense in section 21, and the phrase is also used by Aristotle, but ἐν τῷ μικρῷ is not listed by Bonitz in his Index to Aristotle's works. Perhaps it was originally the name of a place; thus Hill has changed the text to ἐν τῇ Λήμνῳ ("in Lemnos"). Furlanus had already suggested this emendation; he did not put it in his text, but he wrote *in Lemno* in his Latin translation. Wimmer, who gave *in parvo* (*Lemno?*) as his translation, seemed to think that this interpretation might be right. Schneider believed that it would be easy to accept the opinion of Furlanus, but he decided to keep the manuscript reading. Theophrastus has just mentioned Lemnian red ochre, so it is quite possible that he is referring to Lemnos here; but since the correctness of the emendation is not certain, the text has not been changed.

53. *one light-colored.*
The literal meaning of ἔκλευκος is simply "off the white," but its significance in this passage seems clear enough. Apparently Pliny[899] interpreted its meaning in the same way, for he seems to be quoting from Theophrastus at this point when he writes that there were three kinds of red ochre, "the red, the pale red, and one halfway between them."

53. *We call this a self-sufficient kind because it does not have to be mixed, whereas the others do.*
According to this passage the Greek artists intentionally altered the color of natural red ochres when these were not of the proper shade of red. The results of chemical analyses of ancient Greek

[899] XXXV, 31. The text reads: *rubra et minus rubens atque inter has media.*

red ochre pigments appear to indicate that this was their practice. Most of the light-red or pink pigments examined by Midgley[400] contained considerable proportions of calcium carbonate, probably present as an intentional addition, and one such pigment contained both calcium carbonate and calcium sulfate, which is almost certainly a sign of deliberate admixture. Apparently the artists lightened the color of dark-red ochres by mixing them with chalk or other white pigments. Theophrastus also seems to imply that the inferior light-red ochres containing too little ferric oxide were sometimes mixed with ochres containing a higher proportion of ferric oxide in order to produce pigments of the desired depth of color. Attempts to confirm this by chemical analysis have not been successful, as such mixed ochres apparently cannot be distinguished in this way from natural ochres.

53. It is also made by burning yellow ochre, but this is an inferior kind.
It is not necessarily true that red ochre artificially produced by roasting yellow ochre is inferior to natural red ochre, but since most yellow ochres contain a considerable proportion of sand and clay, the final product usually has a lower proportion of ferric oxide and, for this reason, a lower tinting strength when mixed with white pigments. Dioscorides[401] says that the manufactured red ochre is inferior to the kind from Sinope, which, as indicated in the notes on section 52, was probably a natural red ochre containing an unusually high proportion of ferric oxide. Vitruvius[402] states, however, that burnt ochre was very useful for stucco work.

53. Kydias.
Probably this is the artist Kydias mentioned by Pliny,[403] who refers to his most important painting and says that he flourished at the same time as Euphranor, who distinguished himself far beyond all other artists in the hundred and fourth Olympiad.[404] Therefore Kydias' discovery was probably made about half a century before Theophrastus described it. It is uncertain, and perhaps doubtful, whether Kydias was the first to discover how

[400] "Chemical Analysis of Ancient Athenian Pigments," pp. 18-21.
[401] V, 112 (Wellmann ed., V, 96). [402] VII, 9, 2.
[403] XXXV, 128, 130. [404] 364-360 B.C.

to change yellow ochre into red ochre by roasting it, though no definite evidence exists for any earlier discoverer. Undoubtedly the effect of heat on yellow ochre must have been noticed in earlier times, but it is not at all improbable that the artist Kydias was actually the first who used the process deliberately to make a red pigment.

54. *New earthen vessels are covered with clay and placed in ovens; for when the vessels become red-hot, they heat the ochre, and as they become hotter in the fire, they make the color darker and more like glowing charcoal.*
It is not clear from the text whether open or closed pots were used. However, it is unlikely that closed pots were used for roasting yellow ochre since some vent must be provided for the steam that is produced when the ochre is heated. Probably the clay was applied to the exterior of the pots to protect them to some extent from the intense heat of the fire; this procedure would minimize breakage and the consequent loss or contamination of the product. Since new pots were used, it is probable that the vessels often cracked or broke on heating and so could not safely be used again.

The conversion of yellow ochre into an artificial red ochre by roasting is essentially a process of dehydration: the water in the hydrated ferric oxide of the yellow ochre is expelled to form anhydrous ferric oxide of a characteristic red color. Though later ancient writers on technical subjects also allude to this process, they add very little to what Theophrastus says about the procedure employed by the ancient technicians. Pliny[405] merely repeats the account given by Theophrastus, and Dioscorides[406] only mentions the process. Vitruvius[407] does, however, describe very briefly a somewhat different process. He states that burnt ochre was made by heating a clod of good yellow ochre to a glow on a fire and then quenching it in vinegar. This is actually a more primitive process than the one described by Theophrastus much earlier.

Since many technical works composed through the centuries contain accounts of very similar processes for preparing artificial red ochre, it seems that the procedure here described by Theo-

[405] XXXV, 35.   [406] V, 112 (Wellmann ed., V, 96).
[407] VII, 11, 2.

COMMENTARY

phrastus was practiced, in principle at least, more or less continuously from ancient times until a comparatively recent date. Indeed, the manufacture of an artificial red iron oxide pigment by roasting yellow ochre appears to have been the usual method for making such a pigment in European countries until it was first made on a large scale from the waste products of the iron and steel industry. Even at the present time, certain natural earths are roasted in preparing them for use as paint pigments.

55. *There are three kinds of kyanos, the Egyptian, the Scythian, and the Cyprian.*

It was explained in the notes on section 31 that κύανος was a general term denoting both a particular blue precious stone and various blue pigments. In the present section the word obviously refers only to pigments. Actually, as is shown by the results of archaeological excavation, only three stable blue pigments were known at the time of Theophrastus: two of these, azurite and lapis lazuli, were natural; and the third, blue frit, was artificial. The Cyprian *kyanos* was in all probability the native *kyanos* mentioned in section 39, and the same as the *kyanos* of the copper mines mentioned in section 52. If so, it must have been azurite, native blue copper carbonate, which is found in most copper mines and in the mineralized areas in their neighborhood. Dioscorides[408] states that *kyanos* was obtained from the copper mines of Cyprus and also from the sand that is found there in certain hollow places on the sea coast. The Cyprian blue pigment could not have been lapis lazuli, since this does not occur on Cyprus. Azurite for use as a pigment was also obtained from less important copper-mining regions, as is shown by the statements of other writers. For example, the author of *De Mirabilibus Auscultationibus* mentions[409] a mine of *kyanos* on Demonesos, the island of the Chalcedonians. This island was probably the modern Khalki, one of the Prince Islands in the Sea of Marmora, where there are copper minerals and traces of ancient mining operations. The Egyptian *kyanos* was undoubtedly the well-known Egyptian blue frit. By elimination, therefore, the Scythian *kyanos* may be identified as powdered lapis lazuli. This identification receives support from the very

[408] V, 106 (Wellmann ed., V, 91).   [409] Sec. 58.

name given to it, since lapis lazuli does not occur in any of the countries immediately around the Mediterranean but only in countries far to the east. The principal ancient source was apparently in what is now called Badakshan.[410] Whether the deposits of lapis lazuli in Badakshan were within the vaguely defined limits of ancient Scythia is doubtful. The name "Scythian" probably became attached to this particular blue pigment because it was exported to Mediterranean countries by Scythian traders, who had received it from still more distant peoples.

55. *The Egyptian is the best for making pure pigments, the Scythian for those that are more dilute.*
In its usual form Egyptian blue frit has a more intense blue color than powdered lapis lazuli. The difference is not so marked when the size of the particles of the two pigments is about equal, but blue frit apparently was always prepared and used in the form of relatively coarse particles, as the examination of ancient specimens has shown. The color that actually results on grinding it to a very fine powder is a dull bluish-grey, as was determined by an experiment on a specimen of the material found in the excavations at Athens. All other ancient pigments were available and useful only in the form of fine powders, and, because of the great difference in the size of its particles, blue frit could not properly be mixed with other pigments. Any attempt to dilute it with a white pigment such as powdered chalk, in order to apply it as a tempera paint, would have been unsuccessful, for the coarse particles of the frit would settle and only the chalk would remain suspended in the paint.

55. *The Egyptian variety is manufactured.*
Since there is no natural product that has the composition of Egyptian blue frit, it is certain that the Egyptian *kyanos* was always artificially prepared.

Although the composition of this pigment was investigated by chemists as early as the second decade of the last century, it was not until 1889 that its true nature was made known. In that year

---
[410] Cf. the notes on *kyanos* in sec. 31.

COMMENTARY

Fouqué[411] showed that the azure-blue color of the pigment was due to the presence of a single crystalline compound, a calcium copper silicate having the composition represented by the formula $CaCuSi_4O_{10}$. Fouqué was able to reproduce this pigment, but it was not until 1914 that the exact conditions necessary for its formation were determined by the detailed experiments of Laurie, McLintock, and Miles.[412] An account of its manufacture given by Vitruvius[413] indicates the general procedure that the ancients followed in making it.

Though blue frit undoubtedly originated in Egypt and was much used there, its use was actually very widespread, for many examples of the pigment have been found in nearly all the principal centers of ancient civilization bordering on the Mediterranean. Its use in Greece at the time of Theophrastus is certain; in the excavations at Athens specimens have been found both of the pigment itself and of objects colored with it ranging in date from the sixth or fifth century B.C. down into the Roman period.[414] It seems unlikely, however, as is implied by the statements of Theophrastus, that this pigment was manufactured outside Egypt until a comparatively late date. Vitruvius[415] states that the methods of making it were first discovered at Alexandria. This must be incorrect, since Alexandria was founded many centuries after blue frit was first known. He adds that Vestorius afterwards began to manufacture it at Puteoli, and it may well be that this was the place where it was first manufactured outside Egypt.

55. *those who write the history of the kings of Egypt state which king it was who first made fused kyanos in imitation of the natural kind; and they add that kyanos was sent as tribute from Phoenicia and as gifts from other*

---

[411] F. Fouqué, *Comptes Rendus Hebdomadaires des Séances de l'Académie des Sciences* (Paris), CVIII (1889), 325.

[412] A.P. Laurie, W.F.P. McLintock, and F.D. Miles, *Proceedings of the Royal Society of London*, LXXXIX-A (1914), 418-29.

[413] VII, 11, 1.

[414] E. R. Caley, *Hesperia*, XIV (1945), 152-56.

[415] VII, 11, 1.

quarters, and some of it was natural and some had been produced by fire.

Though it cannot be determined exactly when Egyptian blue frit was first discovered, the earliest known specimens belong to the Fourth Dynasty, that is, to the period from 2900 to 2750 B.C.[416] Some inscriptions that have recently been discovered support the statement about the gifts and tribute of *kyanos* that were sent to Egypt, though they do not say that they came from Phoenicia. They say that tribute in the form of both natural and imitation lapis lazuli was sent by certain Mesopotamian rulers to Egyptian kings. In one of these inscriptions, for example, it is recorded that the ruler of Assur sent as tribute to Thothmes III three large lumps of genuine lapis lazuli and three pieces of "blue stone of Babel," which was apparently an imitation.[417] Thothmes III flourished around 1500 B.C., but the Egyptian rulers mentioned in the other inscriptions are later.

55. Those who grind coloring materials say that kyanos[418] *itself makes four colors; the first is formed of the finest particles and is very pale, and the second consists of the largest ones and is very dark.*

Blue frit pigment must be the kind of *kyanos* specifically meant here. Neither azurite nor lapis lazuli yields pigments that differ much in color when they are ground to particles of a different size, but, as has already been indicated, blue frit behaves differently. Some exact measurements of the relation between the color of this pigment and the size of its particles have been made by Peterson.[419] When Theophrastus says that four colors were made from this kind of *kyanos*, he must mean four shades of color, but he is making a mere conventional distinction, since the intensity of the blue color of the pigment decreases continuously as its particles are made smaller. However, it is not unlikely that in

---

[416] Lucas, *Ancient Egyptian Materials and Industries,* p. 394.

[417] B. Meissner, *Babylonien und Assyrien* (Heidelberg, 1920-1925), Vol. I, p. 351.

[418] Eichholz thinks that μέν in τὸν μὲν κύανον is superfluous, and that Σκύθην should be substituted for it. He compares Pliny (N. H. XXXIII, 161): *Scythicum mox diluitur et, cum teritur, in quattuor colores mutatur* (*Classical Review,* LXVI [1952], 144).

[419] C. L. Peterson, "Egyptian Blue and Related Compounds," Master's thesis (The Ohio State University, 1950), p. 39.

COMMENTARY

practice the pigment was sorted into four grades, either because this provided a sufficient variety of shades of blue or because with a larger number of shades it would not be easy to distinguish a particular shade from the one next to it. Theophrastus is correct when he says that the finest pigment was pale and the coarsest was very dark. However, in any process of grinding where the finest is described as the first, strictly speaking the coarsest should be described as the last and not the second, as Theophrastus states. This passage is of considerable interest, as Theophrastus is the first to observe the relation between the color of a vitreous or crystalline material and the size of its particles.

The emendation λευκότατον ("very white," "pale") has been accepted in the text instead of λεπτότατον, which appears in the manuscripts. Turnebus accepted λευκότατον, and was followed by some editors, but Schneider and Wimmer preferred to keep the more difficult reading. It seems probable that the copyist simply made a mistake and wrote λεπτότατον because he had just written λεπτοτάτων. Theophrastus is talking about colors, and a word meaning "very pale" is clearly needed to correspond with μελάντατον, which means "very dark." Since a literal translation of λεπτότατον would be unsuitable, it seems better to accept the emendation λευκότατον in the text.

56. white lead.

It has generally been assumed that the word ψιμύθιον always meant the basic carbonate of lead commonly known as white lead, but Bailey[420] has recently identified the *psimythion* of the Greeks and the corresponding *cerussa* of the Romans as normal or basic lead acetate. However, if attention is paid to the ancient authors who discuss the products derived from lead by the corrosive action of vinegar, it seems certain that these terms were general ones that included both soluble lead acetate and insoluble lead carbonate, and that the particular product depended upon the details of the procedure. According to the procedure described by Theophrastus, the product was washed by decantation, which shows that it must have been insoluble in water and was therefore the carbonate and not the acetate. Bailey and other commentators

[420] *The Elder Pliny's Chapters on Chemical Subjects*, Part II, p. 204.

have been puzzled by the accounts that Theophrastus and later authors have given, because no mention is made of the source of the carbon dioxide necessary for the formation of basic lead carbonate. It has been suggested that either the ancient processes were always performed near fires or some extraneous fermenting material was present, and that the ancient authors simply failed to mention these sources of carbon dioxide. What has been generally overlooked is that the air itself and the water used in the washing operations provided enough carbon dioxide to produce basic lead carbonate by the small-scale processes used by the ancients. This was verified by performing in miniature the process here described by Theophrastus. Lead blocks were supported just above a five per cent acetic acid solution in an open vessel, and the white crusts that were formed on the blocks by corrosion were scraped off at intervals of about ten days. The product was found to be a lead acetate containing only a small proportion of lead carbonate; but when it was dissolved in tap water, a turbidity due to the formation of basic lead carbonate was observed almost immediately, and after the solution had stood for a few hours it absorbed enough carbon dioxide from the air to cause the precipitation of a considerable amount of carbonate. The statements of Theophrastus suggest that, in the process he describes, white lead formed in just this way. Either the successive scrapings of the white crusts were thrown into a mortar along with a little water until all the lead was consumed, and then the total product was ground, washed, and separated, or else the individual scrapings, as they were collected, were ground with water for a considerable time before the product was washed and separated by decantation. Another source of the carbon dioxide may have been the so-called vinegar used in the process. If this was merely a spoiled grape juice undergoing both alcoholic and acetous fermentation, ample carbon dioxide would have been available. That this could have been the main source of the carbon dioxide is suggested by the statements in the next section, where Theophrastus describes how verdigris is prepared by the use of grape-residues and says that the process is similar to the one used for making white lead.

The other ancient writers who describe methods of preparing

COMMENTARY

psimythion and *cerussa* supply additional details or give variations in the general procedure. Vitruvius,[421] in speaking of the manufacture of *cerussa* at Rhodes, states that a layer of twigs was placed in a large jar, vinegar was poured over the twigs, and pieces of lead were placed on top of them. He also states that the jars were provided with lids to prevent evaporation. Pliny[422] gives an outline of two processes: one of them is similar to the process described by Theophrastus, the other is similar to the one given by Vitruvius. Pliny probably obtained his information from these authors. Both Vitruvius and Pliny describe the roasting of white lead to produce red lead, which was also a valuable pigment. Since Theophrastus does not mention this additional step, it seems likely that it was unknown in his day, or at least unknown to him. The most extensive description of the lead corrosion process is given by Dioscorides.[423]

It is interesting to note that the most satisfactory white lead for use as a pigment at the present time is still the kind produced by the action of acetic acid vapor on metallic lead in the presence of air and carbon dioxide, and much of it is made by the so-called Dutch process. In this method, as in the ancient ones, the lead is corroded in small pots containing dilute acetic acid; the essential difference is that in the modern process the operation is performed on a large scale in a closed building, and an abundant supply of carbon dioxide is furnished by some fermenting material such as spent tanbark.

Specimens of white lead used by the Greeks have been discovered by archaeologists and positively identified by chemical analysis. The most extensive discoveries were made by Shear[424] during the excavation of the North Cemetery at Corinth, where small covered bowls in the graves of women and girls were found to contain white lead in the form of pressed cubes, irregular pieces, and loose powder. A representative sample from one of these bowls was examined chemically by Foster[425] and found to be lead carbonate. It is interesting that Dioscorides[426] mentions

[421] VII, 12.   [422] XXXIV, 175-76.
[423] V, 103 (Wellmann ed., V, 88).
[424] *Classical Studies Presented to Edward Capps* (Princeton, 1936), p. 314.
[425] W. Foster, *Journal of Chemical Education*, XI (1934), 225.
[426] V, 103 (Wellmann ed., V, 88).

Corinth as one of the centers of white lead manufacture. Ancient specimens of white lead have been found in Attica and could probably be found elsewhere in Greece. Rhousopoulos[427] mentions a specimen of white lead found in a grave not far from the National Archaeological Museum in Athens which he positively identifies as basic lead carbonate. This, in fact, appears to be the first identification of a specimen of ancient white lead by chemical analysis. At least two other specimens of white lead from graves in Attica have been identified chemically.[428] Since some of these specimens of white lead date from as early as the fifth century B.C., it is clear that the manufacture of this pigment began long before Theophrastus wrote about it.

The fact that white lead is found exclusively in the graves of women and girls and in a particular kind of closed bowl or toilet box shows that it was used as a cosmetic. This is amply confirmed by numerous allusions in the writings of classical authors. There are only a few artificially prepared chemical compounds that were known to the ancients, but this one is mentioned in their writings more often than any other. It is apparently first referred to by Xenophon;[429] in a conversation with Socrates, Ischomachus remarks that, in instructing his wife on her duties, he discourages the use of white lead and other cosmetics because they displease him and are devices that are easily discovered. Similar allusions to the use of white lead as a cosmetic are made by Plautus,[430] Ovid,[431] and Martial.[432] This, in fact, seems to have been the chief use of the product in ancient times, even though its poisonous nature was recognized, but it must be remarked in justice to the ancients that the use of poisonous substances in commercial cosmetic preparations is by no means unusual in modern times.

Pliny[433] lists white lead among the paint pigments, but he states elsewhere[434] that it was not suitable for moist fresco work, a statement which is correct from the chemical point of view, since

---

[427] Diergart, *Beiträge aus der Geschichte der Chemie dem Gedächtnis von Georg W. A. Kahlbaum*, p. 193.
[428] *Classical Studies Presented to Edward Capps*, p. 316; Caley, *Hesperia*, XIV (1945), 153-55.
[429] *Oeconomicus*, X, 7.
[430] *Mostellaria*, 258.
[431] *Medicamina faciei*, 73.
[432] *Epigrammata*, II, 41, 12; VII, 25, 2.
[433] XXXV, 37.
[434] XXXV, 49.

basic lead carbonate is decomposed on contact with slaked lime, and the hydrated lead oxides that are produced darken on exposure to light. Although Theophrastus apparently places it among the paint pigments, white lead appears to have been little used for painting in ancient times. Pliny remarks, however, that *cerussa* was suitable for wax painting but not for mural work. In his researches on the colors used in painting by the ancients, Sir Humphry Davy[435] noted the absence of white lead on painted walls, and its general absence has been noted in later and more extensive investigations. At the present time it is regarded as one of the most important white pigments, and it has had widespread use in spite of attempts by various governments from time to time to restrict or even prohibit its use on account of its toxic nature. Though the ancients did not use it very much as a paint pigment in a direct way, they did use white lead, and probably lead acetate also, for the production of red lead, which was a substitute for the more expensive natural pigments, cinnabar and realgar.

57. Red copper.

The word χαλκός is a general term used to denote both pure copper and its alloys, though it usually refers to bronze. Here the qualifying adjective "red" indicates that unalloyed copper was employed in the process.

57. grape-residues.

Evidently the genitive τρυγός cannot refer to "wine-lees," which is the usual meaning of τρύξ, but in this context means the residues of the grapes that remain after the must has been pressed out of them, or more strictly, perhaps, such residues in a state of acetous fermentation.

Other ancient writers describe processes for the manufacture of verdigris. Pliny[436] gives several methods. In one of these—probably the same as the one given here by Theophrastus—copper was buried in grape skins and the verdigris was scraped off after ten days. In another method perforated pieces of a copper alloy were suspended in closed casks over vinegar. In a third method vinegar was sprinkled over filings of the metal, and the mixture was stirred

---

[435] *Philosophical Transactions of the Royal Society of London*, CV (1815), 97-124.
[436] XXXIV, 110-11.

several times a day with a spatula until all the copper had reacted. Dioscorides[437] also describes similar processes, including one that resembles the method described by Theophrastus. In this process plates of copper were buried in sour grape skins for a number of days.

A method of preparing verdigris by means of acetic acid vapor acting on a sheet of pure copper is described in detail in the *Stockholm Papyrus*,[438] and other accounts appear in medieval technical works.

It is interesting to note that the particular method mentioned here by Theophrastus is, in principle at least, still in use today and probably has been used more or less continuously throughout the intervening centuries. The method of preparing verdigris by the action of sour grape skins on copper plates is mentioned by early modern writers on chemistry and chemical technology. Boerhaave,[439] for example, refers to it, and Hill[440] gives an account of the commercial process that was usual in his time. At present, the manufacture of verdigris by this method is centered in the wine districts of France, particularly at Grenoble and Montpellier.[441] Usually the marc, which is the waste matter consisting of the skins and stems of grapes, is first allowed to ferment, either in large vessels or in special rooms. After the fermentation has proceeded to the proper stage, the pasty mass is spread on thin sheets of copper. Next, piles of alternate sheets of copper and layers of fermented marc are built up. These are allowed to stand from two to five weeks, depending upon the temperature, and are then dismantled. If the process has been successful, the copper sheets are covered with fine green crystals of copper acetate. The sheets are then exposed to the air and moistened from time to time with water or damaged wine. As a result of this treatment they become coated with a thick layer of basic copper acetate which is detached, kneaded with a little water, and pressed into cakes or

---

[437] V, 91 (Wellmann ed., V, 79).
[438] Lagercrantz, *Papyrus Graecus Holmiensis*, pp. 20, 194.
[439] H. Boerhaave, *Elements of Chemistry*, trans. T. Dallowe (London, 1735), Vol. II, p. 152.
[440] *Theophrastus's History of Stones*, p. 134.
[441] T. E. Thorpe, *Dictionary of Applied Chemistry* (London, 1921), Vol. I, p. 24; F. Ullmann, *Enzyklopädie der technischen Chemie* (Berlin, 1929), Vol. IV, p. 676.

COMMENTARY

put in paper containers for the market. Variations in this large-scale process also occur. In some factories, copper turnings are mixed with the fermented marc and, after the reaction has taken place, the verdigris is separated from the copper that remains. The process is also conducted on a small scale by methods which probably differ very little from those employed in ancient times. The blue verdigris produced by this so-called French process consists chiefly of monobasic copper acetate, $Cu(C_2H_3O_2)_2.Cu(OH)_2.5H_2O$.

Theophrastus, like Vitruvius,[442] lists verdigris among the pigments, but Pliny nowhere mentions the use of it in his discussion of painting. Although verdigris has never been found in modern investigations of ancient pigments, this is no proof that it was not used as a pigment. Since basic copper acetate is not a very stable compound, any pigment composed of it would probably have changed completely in the course of the centuries to some more stable copper compound such as the basic carbonate. But the Stockholm Papyrus provides definite evidence of the use of verdigris as a coloring material and lists it as an ingredient in many of the recipes for the preparation of imitation gems. The statements of Pliny[443] indicate that the compound was used extensively by the ancients in the preparation of various remedies.

58. *There is also a natural and a prepared kind of cinnabar.*

Theophrastus has previously shown that the blue pigments and the red iron oxide pigments are both natural and artificial. He seems to think that there are also two kinds of cinnabar, though all the available evidence indicates that this pigment was not produced artificially by the ancients. The earliest mention of the artificial preparation of cinnabar occurs in certain technical recipe books of the Middle Ages.[444] The subsequent statements of Theophrastus in this section show clearly that the real difference between the two kinds was in their mode of occurrence: in some places cinnabar was found in a pure enough state to be used directly, but in others it was mixed with extraneous matter from

[442] VII, 12, 1.   [443] XXXIV, 113-15.
[444] Stillman, *The Story of Early Chemistry*, p. 186.

which it had to be separated. Probably Theophrastus saw little or no difference between the mechanical refining process by which cinnabar was obtained from the crude material and the chemical processes by which the artificial blue or red pigments were produced.

Cinnabar, native mercury sulfide, is first mentioned as a particular kind of stone by Aristotle,[445] but Theophrastus gives the earliest account of it in this section and in the two that follow. However, it was known and used by the Greeks long before the time of Aristotle and Theophrastus. When Rhousopoulos[446] was investigating the pigments on some of the limestone statues of the sixth century B.C. in the Acropolis Museum at Athens, he found that a bright red pigment was native mercury sulfide, and that one of the dark pigments was the same substance altered by exposure to light. Rhousopoulos also found cinnabar present as a coloring material on lecythoi of the fifth century B.C. Traces of it were also found on part of the inside surface and the rim of a small black-glaze bowl of the late fifth century B.C. which was discovered in the excavations at the Athenian Agora.[447] This was apparently a vessel in which the pigment had been mixed for painting. A scallop shell, probably of the third century B.C., which was found in the same excavations contained a small amount of cinnabar and was evidently used for the same purpose. It is not unlikely that cinnabar was used in Asia Minor, and perhaps elsewhere, long before the time of the Greeks. The discovery of a very ancient cinnabar mine near Iconium, which contained stone hammers in the workings, seems to show the earlier use of cinnabar in Asia Minor.[448] There is, however, no evidence of its use in ancient Egypt or in the early civilizations of Mesopotamia.

Though cinnabar is of rather frequent occurrence on Greek objects after the fifth century B.C., it appears to have been used much less frequently than the red iron oxide pigments. Because it was scarce, it was probably always more costly than these other red pigments and was therefore used more sparingly.

[445] *Meteorologica*, III, 6, 378A (26).
[446] Diergart, *Beiträge aus der Geschichte der Chemie dem Gedächtnis von Georg W. A. Kahlbaum*, pp. 180-81.
[447] Caley, *Hesperia*, XIV (1945), 153.
[448] Gowland, *Archaeologia*, LXIX (1917-1918), 157.

COMMENTARY

58. *The cinnabar in Iberia, which is very hard and stony, is natural, and so is the kind found in Colchis.*

Pliny,[449] who is quoting from Theophrastus at this point, translates the Greek place name Ἰβηρίαν (Iberia) as *Hispania* (Spain), and those who have commented on this passage in the *Natural History* seem to have assumed generally that Pliny was correct in his translation. Hill[450] and others after him also translated this place name in the same way; apparently they accepted Pliny's interpretation and knew of the rich cinnabar deposits in Spain that had been exploited as far back as Roman times. Lenz[451] seems to have assumed without question that Spain is the locality to which Theophrastus refers. But it is actually very doubtful whether the cinnabar deposits on the Iberian Peninsula were known, except perhaps locally, as early as the time of Theophrastus, and still more doubtful whether the Greeks obtained cinnabar from that source. The following statements of Vitruvius are important, since they suggest a later time for the discovery, or at least the foreign exploitation, of the cinnabar deposits in Spain:

> It is said that it was first found in the Cilbian districts belonging to the Ephesians .... However, the workshops which were once at the mines of the Ephesians have now been transferred to Rome, because this kind of ore was later discovered in certain districts of Spain. The lumps of ore are brought from the mines there and treated in Rome by public contractors.[452]

From these statements it seems evident that the deposits in Spain were exploited later than those near Ephesos, possibly only after the latter could no longer be worked profitably. Since, according to this account of Theophrastus, the deposits near Ephesos were being worked at the same time as those in the country he calls Iberia, it follows that this Iberia could not have been Spain,

---

[449] XXXIII, 114.
[450] *Theophrastus's History of Stones*, p. 137.
[451] *Mineralogie der alten Griechen und Römer*, p. 26.
[452] VII, 8, 1, and VII, 9, 4. The text reads: "*id autem agris Ephesiorum Cilbianis primum esse memoratur inventum ... quae autem in Ephesiorum metallis fuerunt officinae, nunc traiectae sunt ideo Romam quod id genus venae postea est inventum Hispaniae regionibus, ⟨e⟩ quibus metallis glaebae portantur et per publicanos Romae curantur.*"

as has been generally assumed. Hence Theophrastus probably meant the other ancient country known by the name of Iberia, a country corresponding to the eastern part of the present Transcaucasian Georgia. Colchis, mentioned here along with Iberia, was situated along the eastern shore of the Euxine south of the Caucasus, and was therefore a country corresponding to the western part of the present Georgia. Since Colchis, which is mentioned in the same context, had a common boundary with this eastern Iberia, it is probable that Theophrastus was thinking of this Iberia rather than Spain. The cinnabar may well have been found in a single district common to both these ancient countries.

It is odd that Theophrastus should describe the cinnabar found in Iberia and Colchis as very hard and stony. Cinnabar is not a very hard mineral. Though the crystalline and massive varieties have a hardness of 2 to 2.5 on the Mohs scale—approximately that of rock salt—the earthy varieties are very soft. His statement, therefore, does not appear to be based on actual observation, though perhaps he is merely emphasizing the pronounced difference in hardness between the crystalline and the earthy varieties. This seems a likely explanation, since the impure kind from near Ephesos was probably the earthy variety, whereas the so-called natural kind from Iberia and Colchis was probably a pure crystalline or massive variety.

58. *They say that this is found on cliffs and is brought down by arrows that are shot at it.*
This story is repeated by Pliny[453] with only minor changes. Though it seems to be fabulous, it may, like many other ancient stories, have a real basis. It suggests, at least, that the cinnabar of Colchis was mined in rugged country, and this in turn suggests that the mining district was situated in the northeastern part of Colchis in or near the Caucasus, which is characterized in many places by unusually precipitous rock formations.

58. *The prepared kind comes from one place only, a little above Ephesos.*
That cinnabar actually occurs not far inland from the site of the

[453] XXXIII, 114.

COMMENTARY

great Ionian coastal city of Ephesos has been proved by modern geological exploration. The deposit is located about 65 kilometers southwest of Smyrna, less than 50 kilometers directly inland from the site of Ephesos.[454] Here cinnabar is found scattered through slate and quartz in a vein 15 to 25 meters wide, but the deposit is not considered worth working at the present time. This statement of Theophrastus agrees with the statements of Vitruvius which indicate that the cinnabar mines near Ephesos were finally worked out, so that mining was no longer profitable.

The phrase μικρὸν ἐν καλοῖς, which appears in the text, has not been translated. Schneider thought that it concealed the name of the region mentioned by Vitruvius and Pliny where the Ephesians obtained their cinnabar. Pliny's statement is especially significant, because in the section of his work in which it appears he has apparently taken all his information about cinnabar directly from this part of the treatise *On Stones*. His words are as follows: *optimum vero supra Ephesum Cilbianis agris, harena cocci colorem habente*[455] (the best comes from the Cilbian district above Ephesus, where the sand has the color of scarlet dye). However, it is not at all unlikely that Pliny may have taken his information about the *agri Cilbiani* from the statement of Vitruvius.[456]

If the phrase μικρὸν ἐν καλοῖς is in its right position in the text, it ought to refer to the refining operation and not to the place from which the cinnabar came. If it refers to the place, it should follow ὑπὲρ Ἐφέσου ("above Ephesos"), where the word μικρόν also occurs. The second μικρόν would then be an incorrect repetition of the first, and ἐν καλοῖς, as Schneider suggested, would be a mistake for ἐν ἀγροῖς Κιλβιανοῖς. As it stands in the text, it is impossible to translate it. It was rightly bracketed by Schneider, and Wimmer thought it might be a repetition of ἐν χαλκοῖς ("in copper vessels"), which comes immediately before it in the text.

58. *The washing is done from the top, and separate portions are wetted one after the other; what is left at the*

---

[454] Schmeiszer, *Zeitschrift für praktische Geologie*, XIV (1906), 191.
[455] XXXIII, 114.
[456] VII, 8, 1.

bottom is cinnabar, and the washings are what remains above in larger quantities.

The Greek phrase ἐν πρὸς ἓν ἀλείφοντες is somewhat obscure; the verb would normally mean "anointing," and here it seems to mean "wetting." But it is clear that the process involved successive washings and that it was a method of separating pure cinnabar from its impurities that depended upon the difference in their specific gravities. The specific gravity of pure cinnabar is slightly over eight, whereas the specific gravity of the gangue in which it is usually found is less than three. Therefore, when the crude mineral that had been ground was suspended in water, the cinnabar settled more rapidly than the impurities, which by skillful manipulation could be poured off with the water. The mineral had to be ground thoroughly before it was washed; this released the cinnabar enmeshed in the gangue and reduced all the material to particles of approximately the same size, so that sharp separation would occur on washing.

As the earliest account of the process of separating a pure mineral from its associated impurities, this description of the method used at the cinnabar mines near Ephesos is of considerable historical interest.

59. *They say that Kallias, an Athenian from the silver mines, discovered and demonstrated the method of preparation.*

Pliny,[457] who is quoting Theophrastus at this point, makes it appear that Kallias' discovery was made at the silver mines in Attica, and apparently because of this incorrect quotation some have assumed that the process of separating pure cinnabar from the crude ore was discovered at the Laurion silver mines, or that it was discovered there and then used for the treatment of local ore. But neither the wording of this passage in Theophrastus nor the geological facts warrant either of these assumptions. Cinnabar does not occur now in the Laurion mining district, nor is there any evidence that it ever did occur there. Moreover, Theophrastus is obviously still speaking of the refining process used at the cinnabar mines near Ephesos. It is far more likely that Kallias, as

[457] XXXIII, 113.

COMMENTARY

an expert on the treatment of ores at the silver mines at Laurion, went to the cinnabar mines near Ephesos to devise a process of treating the ore. It is obvious that lead-silver ores were concentrated at Laurion by a skillful process of washing; the remains of ancient washing tables and other apparatus can still be seen there. Furthermore, the archaeological evidence shows that the refining operations at Laurion were conducted on a larger and more elaborate scale than those at any other Greek mining site. It is therefore to be expected that mine operators at other places, faced with the necessity of treating unusual ores by washing, might have called upon an experienced man from Laurion to devise a suitable process.

Kirchner[458] and Jaeger[459] both suggest that the Kallias mentioned here was the son of Hipponicos, one of the wealthy Athenian operators of the silver mines at Laurion. The date is suitable, and there is also the evidence of Xenophon.[460] But Kallias was a common Greek name, and the evidence as a whole is so scanty that there can be no certainty about this identification.

59. *for thinking that the sand contained gold because it shone brightly, he collected it and worked on it. But when he saw that it did not contain any gold, he admired the beauty of the sand because of its color and so discovered this method of preparation.*

This sounds very much like the usual ancient story invented after some important event in order to explain it. Though crystalline cinnabar glistens in the light, the more common earthy varieties of this mineral do not, and there is nothing about the luster of crystalline cinnabar or the color of any variety of this mineral that would lead anyone to suspect that it might contain gold. However, cinnabar is sometimes associated with pyrite, the so-called fool's gold; and if the "sand" or cinnabar ore investigated by Kallias did contain some pyrite, he might easily have been misled by its luster and color. If he found pyrite in the ore, he might have paid little attention to the striking color of the cinnabar

---

[458] J. Kirchner, *Prosopographia Attica* (Berlin, 1901), Vol. I, p. 521.
[459] W. W. Jaeger, *Diokles von Karystos* (Berlin, 1938), p. 120, footnote.
[460] *Symposium*, I, 2; *De Vectigalibus*, IV, 15.

until his attempt to obtain gold had ended in failure. Thus it is possible that the story may have some truth in it.

Pliny[461] relates a somewhat analogous story about an attempt to obtain gold from orpiment, the golden-yellow, native sulfide of arsenic. This is said to have been ordered by the Emperor Caligula. According to Pliny, a small quantity of gold was actually obtained in this experiment, but the amount was so small that the attempt was considered a failure. Probably many experiments of this sort were tried by the ancients.

59. *This did not happen long ago, but about ninety years before Praxiboulos was archon at Athens.*
This interpretation agrees with that of Pliny, who paraphrases the passage as follows: *Theophrastus LXXXX annis ante Praxibulum Atheniensium magistratum—quod tempus exit in urbis nostrae CCCXLVIIII annum—tradit inventum minium a Callia Atheniense* ....[462] (Theophrastus states that ninety years before Praxibulus was archon of the Athenians—a date that corresponds to the 349th year of our City—cinnabar was discovered by Callias the Athenian ....) Thorndike[463] explained the passage in quite a different way, for he believed that the preposition εἰς in Theophrastus meant "back to" rather than "prior to," and therefore he placed the time of the discovery in the archonship of Praxiboulos. In his opinion this was also the interpretation of Hill, though Hill's translation of the passage is actually so ambiguous that it may be taken either way. It reads as follows, "And this is no old thing, the invention being only of about ninety years date; Praxibulus being at this time in the Government of Athens."[464] Since Praxiboulos is known to have been archon in 315-314 B.C., the interpretation of Thorndike implies that this treatise was composed in 225-224 B.C., sixty years after the death of Theophrastus. Thorndike therefore suggested that the treatise was written by someone else and ascribed to Theophrastus. However, the text does not

[461] XXXIII, 79.
[462] XXXIII, 113. See also Bailey, *The Elder Pliny's Chapters on Chemical Subjects*, Part I, pp. 118-21, 218.
[463] C. Singer and H. E. Sigerist, *Essays on the History of Medicine Presented to Karl Sudhoff on the Occasion of His Seventieth Birthday* (Oxford, 1923), pp. 73-74.
[464] *Theophrastus's History of Stones*, p. 139.

COMMENTARY

support Thorndike's views. Examples of a similar use of the preposition εἰς in both Thucydides[465] and Theophrastus[466] show that the present interpretation is correct. The passage in the *History of Plants* is especially important because of the similarity in construction and lack of ambiguity. It reads as follows: φασὶ δ' οἱ Κυρηναῖοι φανῆναι τὸ σίλφιον ἔτεσι πρότερον ἢ αὐτοὶ τὴν πόλιν ᾤκησαν ἑπτά· οἰκοῦσι δὲ μάλιστα περὶ τριακόσια εἰς Σιμωνίδην ἄρχοντα Ἀθήνῃσιν. (The people of Cyrene say that silphium—laserwort—first appeared seven years before they founded their city; now they have been living there about three hundred years prior to the archonship of Simonides at Athens.) Since Simonides was archon in 311-310 B.C., the meaning ascribed to εἰς by Thorndike is impossible here. Theophrastus apparently used the date of the archonship of Praxiboulos, 315-314 B.C., as a reference point, because he was writing in that year; and if this is so, the discovery of Kallias was made about 405-404 B.C., which would have been ninety years before.

Jaeger[467] has recently advanced a theory that the treatise *On Stones* was not composed in the year 315-314 B.C., as seems to be indicated in this section, but twenty years or so later. In section 28 Theophrastus mentions a certain Diokles, whom Jaeger identifies as Diokles of Karystos, a noted Athenian physician who flourished in the fourth century B.C. In this Jaeger agrees with previous views, but he also thinks that Diokles was still alive in the opening years of the third century, since there appears to be evidence that Diokles wrote a letter to Antigonus, King of Macedon, sometime between the years 305 and 301 B.C., and wrote a medical work shortly after 300 B.C. Moreover, Jaeger believes that by the use of the imperfect (Διοκλῆς ἔλεγεν) in section 28, Theophrastus shows that Diokles was no longer alive. But he also implies that Diokles was his contemporary and that he had heard him speak. Thus it seems to follow that Theophrastus composed his treatise *On Stones* sometime between the opening years of the third century and the time of his death about 287 B.C. Though this theory rests to some extent on conjecture, it does supply a possible explanation of the peculiar method of dating used in this

[465] I, 13, 3. [466] *History of Plants*, VI, 3, 3.
[467] *Diokles von Karystos*, pp. 1-5, 114-23.

· 201 ·

section of the treatise *On Stones*. Jaeger's explanation is that Theophrastus obtained his information from another author who wrote during the archonship of Praxiboulos and stated that the discovery had been made about ninety years before; then Theophrastus, as an easy and convenient way of dating the discovery, added the statement about the archonship without changing the wording of the passage from which he took his information. He could easily have taken it from the report of an assistant, as he appears to have followed this practice freely in composing his *History of Plants*. The use of the verb φασί ("they say") at the beginning of this section definitely supports this explanation, as it shows that he is depending upon the authority of someone else for his information.

Whatever may be the true explanation of this method of dating by the year of the archon, it has the merit of fixing both the time of Kallias' discovery and the time at which a record was made of it by Theophrastus himself or by some contemporary upon whom he depended for his information. It really matters little whether 315-314 B.C. or the slightly later date suggested by Jaeger is accepted as the date of the composition of the treatise, since both fall within the known lifetime of Theophrastus and both are of equal value as an indication that he was the actual author.

60. *paints*.

The word ἄλπεις, which occurs in the manuscripts and the text of Aldus, cannot be translated. Four of the editors, including Hill, read ἀλιπεῖς; apparently they are referring to the kind of earth that is not greasy and is suitable for painting (sec. 62). Since a word meaning "paints" seems to be needed, one possibility is ἀλιφάς, a late spelling of ἀλοιφάς, the noun derived from ἀλείφω ("to anoint or paint"). The word occurs in the singular (ἀλιφήν) in an inscription of the third century B.C., which coincides with part of the lifetime of Theophrastus. This inscription, which refers to the materials for the erection of a temple at Eleusis, reads as follows: καὶ κόλλαν ὠμοβόϊον καὶ τἆλλα ὧν ἂν δέῃ πρὸς τὴν ἐργ⟨ασίαν καὶ τὴν ἔρ⟩εψιν καὶ τὴν ἀλιφὴν τῶν ξύλων

COMMENTARY

καὶ τὴν δόρωσιν . . . .⁴⁶⁸ (and raw ox-glue and other things necessary for the work, for the roofing, the painting of the wood, and the plastering . . . .)

60. *and others for both purposes equally, such as quicksilver; for this has its use too.*
Since cinnabar and mercury often occur together, it is likely that mercury was known as early as cinnabar; and, as archaeological discoveries have shown, cinnabar was undoubtedly known in Greece by the sixth century B.C. and in Asia Minor probably much earlier.[469] That cinnabar and mercury occurred together in at least some ancient deposits is clear from a statement of Vitruvius,[470] who says that in the mining of cinnabar this mineral shed tears of quicksilver under the blows of the tools and that these tears were at once gathered by the diggers. Free mercury actually occurs in the cinnabar deposits that still exist in the district near Ephesos, where the ancient deposits were found.[471] However, there is no archaeological or literary evidence that the Greeks knew about mercury as early as the sixth century. It is first mentioned by writers of the fourth century B.C., and the earliest allusion to it seems to have been made by the comic dramatist Philippos; according to Aristotle[472] he explained the movements of a wooden statue of Aphrodite by saying that the sculptor Daedalos poured quicksilver into it. Theophrastus is the first to describe the preparation of mercury from cinnabar and the first to mention that mercury had some practical use.

Vitruvius, who wrote in the first century B.C., is the earliest ancient author to give detailed information about the practical uses of mercury in antiquity. This is what he says:

> Moreover, it is convenient to use it for many purposes. In fact, neither silver nor copper can be gilded properly without it. And when gold has been woven into a garment, and the garment becomes worn with age so that it is no longer respectable to use, the pieces of cloth are put into earthenware pots and burned up over a fire. The ashes are then thrown into water and quicksilver

[468] *Supplementum Epigraphicum Graecum*, III, 147.
[469] Cf. the notes on sec. 58.   [470] VII, 8, 1.
[471] Schmeiszer, *Zeitschrift für praktische Geologie*, XIV (1906), 191.
[472] *De Anima*, I, 3, 406B.

added to them. This attracts all the little pieces of gold and makes them combine with itself. When the water has been poured off, the residue is emptied into a cloth and then squeezed in the hands; the quicksilver, because it is a liquid, escapes through the loose texture of the cloth; the gold, brought together by the squeezing, is found inside in a pure state.[473]

Pliny[474] also describes the practical uses of quicksilver, but he adds little to what Vitruvius says. The *Leyden Papyrus X* contains recipes for gilding with the aid of mercury, for preparing gold amalgams for lettering in gold, for "silvering" copper objects with mercury, and for making various simple and complex amalgams of base metals in imitation of silver.

60. *It is made when cinnabar mixed with vinegar is ground in a copper vessel with a pestle made of copper.*

A lacuna in the manuscripts and Aldus shows that a word is missing before τριφθῇ. This must be κιννάβαρι, which can be supplied from Pliny,[475] who seems to be quoting from Theophrastus. The chemical facts also require it.

This was no mere mechanical method for the liberation of the metal from a natural mixture of mercury and cinnabar, but it was a true chemical process that depended upon the displacement of the mercury from the cinnabar by the more active metal placed in contact with it. Lenz[476] doubted that cinnabar would be decomposed by the process here described by Theophrastus, and Blümner[477] apparently accepts this opinion, but Bailey[478] has demonstrated by an experiment that cinnabar can be decomposed when it is subjected to this treatment. This experiment was performed by grinding cinnabar with copper turnings and vinegar. Though the reaction was found to proceed very slowly when the mixture was cold, it took place readily enough when it was warmed, and the products were copper sulfide and mercury. However, the liberated mercury soon united with some of the unchanged copper to form an amalgam of copper and mercury.

[473] VII, 8, 4. [474] XXXIII, 99, 125. [475] XXXIII, 123.
[476] *Mineralogie der alten Griechen und Römer*, p. 26.
[477] *Technologie und Terminologie der Gewerbe und Künste bei Griechen und Römern*, Vol. IV, p. 98.
[478] *The Elder Pliny's Chapters on Chemical Subjects*, Part I, p. 223.

COMMENTARY

In order that pure mercury may be obtained by this method, an additional operation is evidently needed: namely, the heating of the amalgam and the condensation of the pure mercury volatilized by the heat. Thus the process described by Theophrastus produced an amalgam and not pure mercury. Possibly such impure mercury was all that was produced and used at the time of Theophrastus, though it seems more likely that the amalgam was distilled in order to obtain the pure metal. The separation of mercury by distillation was certainly known a little later in ancient times, since both Dioscorides[479] and Pliny[480] describe its isolation by this method. Since Theophrastus does not mention that the mixture ought to be heated and a simple distillation performed, it is probable that he did not know all the details of the process used in his day.

This passage is not only the first account of the isolation of mercury from cinnabar but also the earliest description of any method of isolating a metal from one of its compounds.

60. *And perhaps several other things of this kind could be discovered.*
This statement seems to imply that Theophrastus was in favor of experiment, though perhaps it only shows that he thought that such technical processes were discovered by chance. It certainly does not show that he appreciated the importance of systematic experiment.

61. *Among the substances obtained by mining there still remain those that are found in earth-pits.*
This section serves as an introduction to the remainder of the treatise, which deals principally with the earthy minerals or with products derived from them.

The fundamental meaning of γεωφανής is "looking like earth." This suggests that Theophrastus may have been referring to substances "that have the appearance of earths." However, Liddell and Scott's lexicon states that in this particular passage the word means "a spot where a kind of ochre was dug." It seems clear that this definition has been taken from the translation given by

[479] V, 110 (Wellmann ed., V, 95).   [480] XXXIII, 123.

Hill, who speaks of "remarkable earths dug out of pits." Though the original meaning is attractive, the use of the preposition ἐν does seem to suggest a place, and for this reason the translation "earth-pits" has been adopted.[481]

61. *And all sorts of colors are obtained from them owing to the differences of the matter they contain* . . . . The phrase immediately preceding this corrupt passage refers to substances caused by "some conflux and separation of matter which is purer and more uniform than that of the other kinds"; it is a recapitulation of what was said before, especially in section 2. This passage appears to be a similar recapitulation of what was said at the beginning of section 50. Though it is impossible to emend this passage with any certainty, what Theophrastus says in section 50 about the differences in savors may supply a hint. Turnebus changed οὔντων to ποιούντων, and four of the editors, including Hill, followed him. He also inserted καί after ὑποκειμένων. Schneider suggested καὶ διηθούντων, referring to matter "filtering through" or "percolating."

61. *some of them are softened and others are ground and melted, and in this way the stones that are brought from Asia are constructed.*
That μαλάττοντες ("softening") really means fusing or sintering in this context, and that τήκοντες ("melting") means dissolving or leaching, seems more than likely from the use made of these words by Aristotle and Theophrastus.[482] Though the word order suggests that the material was ground after it was dissolved or leached, it is highly probable that the grinding was done first. If so, Theophrastus is merely reversing the natural order of words, following the construction known in Greek as "hysteron proteron," just as he did in section 58 (πλύνουσι καὶ τρίβουσιν).
Actually μαλάττοντες is Schneider's emendation for μελαντώντες, which appears in the manuscripts and makes no sense. But the emendation seems to be correct, especially as a contrast is again

---

[481] The *Etymologicum Magnum* includes the noun γεωφανεῖον, which it describes as a place where there is a mine of earth (χωρίον ἐστὶν ἐν ᾧ γῆς εἶναι μέταλλον).
[482] Cf. the notes on sec. 48.

COMMENTARY

made between the verbs τήκω and μαλάττω, which were used in section 48.

Celsus,[488] Dioscorides,[484] and Galen,[485] when they discuss the sources and properties of certain mineral drugs, mention an Asian or Assian stone, which apparently came from the town of Assus in the Troad. They also speak of the "flower" of this stone, which appears to have been a natural efflorescence that formed on it. Though Theophrastus says that the "stones" were brought from Asia, it is not at all likely that they were the same as this stone of Assus, for this appears to have been a natural product and the so-called "stones" were evidently artificial. But it is difficult to determine what kind of artificial product they were. Since they were made by heating, grinding, and leaching earthy minerals, it is most likely that they were crystals of salts, such as alum, copperas, and blue vitriol, prepared from the impure natural sulfates. Dioscorides,[486] Pliny,[487] and Galen[488] clearly show that such crystallized salts were made in ancient times. Galen actually visited a factory on the island of Cyprus where he saw one of these salts being produced; this was probably hydrated copper sulfate contaminated with iron sulfate. He describes how the green solution containing the dissolved salt was drained into a warm cave where the salt was allowed to crystallize.

Though the Asian or Assian stone and these "stones" from Asia were in all probability not the same, there may have been a relationship between them. Descriptions of the "flower" of this stone given by Dioscorides[489] and by Pliny[490] suggest that this product was the sort of saline efflorescence that occurs on such sulfide minerals as marcasite or pyrite when they become weathered. In other words, the so-called "flower" of Asian or Assian stone may have consisted of natural earthy sulfates, whereas the "stones" mentioned by Theophrastus in this passage may have been such sulfates in the form of purified crystals.

[483] IV, 31 (Daremberg ed.).   [484] V, 141 (Wellmann ed., V, 124).
[485] *De simplicium medicamentorum temperamentis ac facultatibus*, IX (Kühn ed., XII, 194, 202).
[486] V, 114 (Wellmann ed., V, 98).   [487] XXXIV, 123-25; XXXV, 183-88.
[488] *De simplicium medicamentorum temperamentis ac facultatibus*, IX (Kühn ed., XII, 238-41).
[489] V, 141 (Wellmann ed., V, 124).   [490] XXXVI, 133.

THEOPHRASTUS ON STONES

62. *The natural kinds of earth, which are useful as well as superior in quality, are three or four in number, the Melian, the Kimolian, the Samian, and a fourth in addition to these, the Tymphaic or gypsos.*

Melian earth, which was found in Melos, an island in the Cyclades, is mentioned by several other ancient writers. Most of them say or imply that it was white, but Dioscorides[491] states that it was ash-colored like the Eretrian earth. This suggests that it probably occurred both in a white and a greyish form. For use as a paint pigment it is probable that only the white form was suitable. Later in this section Theophrastus compares it with Samian earth and gives the impression that it felt rough when it was touched. Dioscorides specifically says that it was rough, and that when it was rubbed between the fingers it made a sound, just as pumice does when it is rubbed. Various conjectures have been made about its identity. Hill[492] described it as a fine white marl, though he gave no reason for this. Lenz[493] thought that Theophrastus was describing a clay, or a chalky clay, and various writers have identified Melian earth as a white clay, though they do not seem to have had any actual knowledge of the kinds of earthy minerals that occur on the island of Melos. However, Stephanides[494] did know that deposits of kaolin occur on Melos, and he also identified Melian earth as a pure white clay or kaolin. But there is an objection to this identification, since the statements of both Theophrastus and Dioscorides plainly show that this earth was not unctuous like clay, but harsh or rough to the touch. Though some impure clays feel rough because of the presence of sand, it seems more likely that Melian earth was not a clay at all but some other substance such as a siliceous earth. What makes this very probable is the actual occurrence of a white siliceous earth on Melos, where several large deposits have recently been discovered and exploited. This is a brilliant white earth that occurs in chalklike masses. It consists of nearly pure silica in a cryptocrystalline form. Though it is soft and extremely fine after the

---
[491] V, 179 (Wellmann ed., V, 159).
[492] *Theophrastus's History of Stones,* p. 142.
[493] *Mineralogie der alten Griechen und Römer,* p. 27.
[494] *The Mineralogy of Theophrastus,* p. 93.

COMMENTARY

usual grinding and washing, it has distinct abrasive properties and can be used as a polishing powder and as a filler in paints.[495] It has appeared on the market under the trade name "Milowite," and the modern exploiters of this siliceous earth seem to regard it as a new mineral product, though it is probably the same as the Melian earth that was so widely known in ancient times.

Kimolian earth took its name from the island in the Cyclades upon which it was found. Kimolos is a small island very close to Melos. Since Theophrastus only names this earth without giving any description, it cannot be identified from what he says. However, the descriptions of Kimolian earth given by later ancient writers such as Dioscorides[496] and Pliny[497] do identify it. Dioscorides states that Kimolian earth occurred in two varieties, white and reddish (literally, inclining to purple), and that it had a certain natural fatness. Pliny mentions its use for cleaning clothes, which at once suggests that it had the properties of a fuller's earth. The principal clayey material of this kind that occurs on the island of Kimolos is a particular variety of sepiolite, a hydrated magnesium silicate to which modern mineralogists have given the name *cimolite*. This has been used in modern times for cleaning cloth, and it is in all probability the same as the Kimolian earth mentioned here by Theophrastus.

The description given by Theophrastus suggests that Samian earth, which takes its name from the island of Samos off the coast of Asia Minor, was in all probability kaolin, hydrated aluminum silicate, or a clay composed mostly of kaolin. Deposits of kaolin and fine clays occur on the island, and in ancient times they were extensively used for the manufacture of ceramic ware. In section 63 there is another indication that Samian earth was kaolin, and this is discussed in the notes on that section. The descriptions given by authors who lived later than Theophrastus further support this identification. For example, Dioscorides[498] mentions that Samian earth clings strongly to the tongue, a special property more or less characteristic of kaolin and of clays that are largely composed of it.

[495] J. N. Wilson, *Chemical Trade Journal*, XCVII (1935), 28; *Sands, Clays, and Minerals*, II, No. 3 (1935), 127-30.
[496] V, 175 (Wellmann ed., V, 156).   [497] XXXV, 195-96.
[498] V, 171 (Wellmann ed., V, 153).

The fourth kind of earth that Theophrastus mentions is the kind that takes its name from Tymphaia, a district in northern Epirus; he seems uncertain whether to classify it as another kind of native earth or as *gypsos*. The same uncertainty can be seen in section 63, where he says that the people near Mt. Athos use the Tymphaic earth for clothes and call it *gypsos*. On the other hand, he states in section 64 that *gypsos* occurs in Tymphaia, so that he probably regarded Tymphaic earth as identical with *gypsos*. Therefore, it would seem that Tymphaic earth was either a natural kind of *gypsos* or a substance that closely resembled one of the minerals included under this term. It may have been an earthy gypsum or a white chalk.

62. *gypsos*.

Although the English word gypsum is derived from the Greek word γύψος, which has often been translated in this way, the descriptions given by Theophrastus in the rest of the treatise show clearly that, in his day at least, the Greek word had a much broader significance. It is, however, certain from some of these descriptions that gypsum, natural hydrated calcium sulfate, was included under the name *gypsos*. On the other hand, it is equally certain that the Greek term was also applied to the artificial partly dehydrated calcium sulfate now known as plaster of Paris. Moreover, some of these descriptions show beyond doubt that *gypsos* must have included a very different substance, our present quicklime or calcium oxide. It apparently also included various preparations made from these different substances. The evidence for the use of the word is discussed in the notes on the remaining sections. Theophrastus not only fails to make distinctions between the different chemical substances included under the term *gypsos*, but he often confuses one with another.

62. *Painters use only the Melian kind; they do not use the Samian, even though it is beautiful, because it is greasy, dense, and smooth.*

Samian earth is not mentioned as a paint pigment by other ancient authors, and it is easy to understand why a material of this nature was not used. There are, however, many allusions in the

COMMENTARY

works of these authors to the other uses of Samian earth, though Theophrastus says little about them. It appears to have been widely used for medicinal purposes, principally as an ingredient of plasters and salves.

62. *For the kind which is . . . and . . . , and is not greasy is more suitable for painting.*
Since a contrast between the properties of Melian and Samian earth is clearly intended, one of these adjectives describing Melian earth should suggest lightness or lack of density. Turnebus substituted ἀραιόν for ἤρεμον; this means "of loose texture" and is the opposite of πυκνός ("compact"). Furlanus and three other editors accepted ἀραιόν but, oddly enough, kept ἤρεμον as well. This word usually means "quiet"; it is not clear why Liddell and Scott's lexicon gives "smoothness" as the meaning in this passage. Perhaps the adverb ἠρέμα was written here, as Turnebus seems to imply by his Latin translation *leniter aequabile*. The second adjective might be the opposite of λεῖος ("smooth"), for smoothness was one of the qualities of Samian earth. Only the ending (δες) of this adjective remains. Turnebus added two syllables and read τραχώδες ("rough"), the opposite of smooth. He was followed by four other editors, but Schneider and Wimmer did not attempt to emend the text. Since it is impossible to know what adjective was used, no emendation has been made.

62. *and the Melian kind has this quality . . . .*
The words τῷ φαρίδι appear in the Aldine text; the manuscripts have the same reading, but the last vowel is elided before the first word of the next section. The text seems to be corrupt, as the meaning of the word φαρίδι is unknown. Heinsius, De Laet, and Hill print it with a capital letter, as if Pharis were the name of a place, and add the preposition ἐν. Thus Hill gives the following free translation, "all which properties the Melian, particularly that of Pharis, possesses." However, no evidence exists that a town or other locality of this name ever existed on the island of Melos. It is significant that neither Schneider nor Wimmer accepted this emendation. Since Turnebus uses the words *suapte friabilitate* in his Latin translation, Schneider suggests ἐν τῷ ψαφαρῷ

or σὺν τῇ ψαφαρότητι. This would mean that the earth is "liable to crumble." Wimmer does not attempt to give the meaning of φαρίδι in his Latin translation. Since Theophrastus speaks in the next sentence of differences (διαφοραί) in Melian and Samian earth, it is possible that the words τῷ φαρίδι conceal the adjective διάφορον, which would agree with the relative ὅπερ and would refer to the distinctive quality of Melian earth. It might be translated as "in marked degree."

63. *The innermost earth is called "the star."*

The wording of the text suggests that the inner stratum was called "the star," but Dioscorides,[499] Pliny,[500] and Galen[501] all show that the name was really given to one kind of Samian earth. Since a lacuna precedes this passage, the missing words may have explained this more clearly. These later writers name κολλούριον or *collyrium* as another variety of Samian earth, and it is possible that Theophrastus also named this other variety in the phrase that originally preceded the present passage. Since the so-called "star earth" was taken from the innermost part of the vein, presumably the other variety came from the outer parts.

The name *collyrium* probably signified that the second kind of Samian earth was shaped in the form of small loaves or rolls, but the meaning of the name given to the first kind is obscure If this explanation of *collyrium* is correct, perhaps the name *aster* was used because this kind was formed into star-shaped cakes. It is also possible that it was used because a star-shaped trademark was stamped on the cakes. The second explanation seems plausible, as another famous earth, the Lemnian medicinal earth, was sold in the form of small cakes stamped with a characteristic trademark.[502] Hill says that "the white was the *Aster*, supposed by many to be a talc, and so called for its shining,"[503] but this explanation of the name is not very likely to be correct. Bailey says that "Dana, no doubt correctly, identifies Samian earth with kaolinite," and he adds that "this sometimes occurs in pearly, hexago-

---
[499] V, 171 (Wellmann ed., V, 153).   [500] XXXV, 191.
[501] *De simplicium medicamentorum temperamentis ac facultatibus*, IX (Kühn ed., XII, 181).
[502] Cf. the notes on sec. 52.
[503] *Theophrastus's History of Stones*, p. 146.

nal plates, often grouped in fan-shaped forms (*aster*), but more often as a clay-like mass, white, grey, or yellow (*collyrium*)."[504] Dioscorides[505] indicates that the variety called *aster* had a lamellar structure, which tends to support the identification suggested by Bailey. It is also possible that the name was used in a metaphorical sense to denote the best variety of Samian earth, which was taken from the middle of the vein. Though some of these explanations are plausible enough, it is impossible to determine with certainty how this name originated.

64. *This earth is used mainly or solely for clothes.*
Theophrastus is the only ancient writer who mentions that Samian earth was used for cleaning or whitening clothes. If this earth was kaolin, as seems probable, it could have hidden the dirt on the surface but not bleached the cloth or actually removed much dirt.

64. *The Tymphaic earth is also used for clothes and is called gypsos by the people who live near Mt. Athos and those districts.*
Tymphaic earth, which was discussed in the notes on section 62, was probably earthy gypsum or chalk, and this may have had the same effect as Samian earth when it was used on clothes; for the material would be impregnated with fine particles of white pigment and so would appear to be clean.

Mt. Athos, which is also known today as the Holy Mountain because of its monasteries, is the most eastern of the three promontories that form the Chalcidic peninsula in the Northern Aegean.

64. *Gypsos occurs in large quantities in Cyprus and can easily be seen; for only a little soil is removed when it is dug up.*
Here the term *gypsos* seems to mean native gypsum. This mineral, some of it in the form of alabaster, is abundant in several places on the island of Cyprus, and at present both untreated gypsum and plaster of Paris are important exports from the island. Sometimes this term denoted alabaster, as is indicated in the notes on section 65.

[504] *The Elder Pliny's Chapters on Chemical Subjects*, Part II, p. 240.
[505] V, 171 (Wellmann ed., V, 153).

64. In Phoenicia and Syria it is made by burning stones, and this also happens in Thourioi.
Here the material is obviously not a natural mineral substance, and the information given by Theophrastus in section 69 identifies it with certainty as quicklime. For further comments see the notes on section 69.

Thourioi was a city of Magna Graecia on the Tarentine Gulf in southern Italy.

64. it occurs in Tymphaia and in Perrhaibia.
It has already been shown that the kind of earth or *gypsos* that occurred in Tymphaia, a district in northern Epirus, was probably earthy gypsum or chalk. Since Perrhaibia was a neighboring district in northern Thessaly, it is likely that the same mineral occurred there.

65. Its nature is peculiar; for it is more like stone than earth, and the stone resembles alabastrites. It is not cut out in a large mass but in small pieces.
Alabastrites was the name specifically applied to Egyptian onyx marble, as is explained in the notes on section 6; at the present time this is sometimes called "oriental alabaster" to distinguish it from ordinary alabaster, which is a form of gypsum. Since it closely resembles Egyptian onyx marble in appearance, it seems probable that the mineral substance which Theophrastus mentions here was in fact this particular variety of gypsum. At any rate, the allusion is certainly to natural gypsum, not to the dehydrated mineral or to any other sort of artificial product.

65. Its stickiness and heat, when it is wet, are remarkable.
There seems to be an inconsistency here. Since no natural mineral substance generates heat to an appreciable extent on being treated with water, it looks as if Theophrastus were now describing an artificial mineral product. The preceding passage suggests that it was partly dehydrated gypsum, but from what follows it seems to have been quicklime. Such inconsistencies not only indicate that different substances were included under the term *gypsos* but

COMMENTARY

they also show the confusion that was caused when the ancients failed to see that each of these substances had distinctive properties of its own. However, the confusion in this treatise may also have arisen because Theophrastus had no first-hand information.

65. *it is used on buildings and is poured around the stone or anything else of this kind that one wishes to fasten.*
Here *gypsos* appears to mean a prepared mortar, and unless this statement applies to a very dry country like Egypt, the material must be lime mortar and not gypsum mortar, since the latter soon disintegrates in wet weather. Though only a few chemical analyses have been made of ancient Greek mortars, they indicate that lime mortar was the only kind used in Greece at the time of Theophrastus.[506]

66. *After it has been pulverized and water has been poured on it, it is stirred with wooden sticks; for this cannot be done by hand because of the heat.*
Though both quicklime and dehydrated gypsum generate heat when mixed with water, quicklime generates far more heat. Since Theophrastus makes a point of mentioning the heat, it is likely that he is referring to mortar made from quicklime and not to gypsum mortar. It is curious that Theophrastus says nothing about the addition of sand or any other filler, since a satisfactory mortar could not have been made without this. Foster's analyses[507] show that the Greeks added about one part of sand to two parts of lime in the preparation of their mortars. Since Theophrastus says nothing about any filler, it is probable that he had no first-hand knowledge of the subject.

66. *And it is wetted immediately before it is used; for if this is done a short time before, it quickly hardens and it is impossible to divide it.*
This sentence differs sharply in significance from the two that precede it, for it indicates that Theophrastus is now speaking of gypsum plaster or mortar, which, in contrast to a lime preparation,

[506] Foster, *Journal of Chemical Education*, XI (1934), 223-24.
[507] *Loc. cit.*

does harden quickly. This is another sign that Theophrastus confused the different materials grouped under the term *gypsos*.

67. *And it can even be removed and calcined and made fit for use again and again.*

Since Theophrastus is alluding to the mortar used in constructing the walls of buildings, the material which was reburnt and used again was almost certainly lime mortar, not gypsum mortar. Both this passage and the preceding section show that the term *gypsos* was used to describe hardened mortar and not simply the essential ingredient of mortar.

67. *but in Italy it is also used for treating wine.*

The emendation τὸν οἶνον ("wine") appears in the text of Turnebus. The manuscripts and the Aldine edition have τὸν οἰκεῖον, which is hard to translate. Hill followed a suggestion of Salmasius and changed this to τὴν κονίασιν ("plastering"), but both Schneider and Wimmer accept τὸν οἶνον.

The accounts of later writers show that the *gypsos* used for treating wine was either lime or partly dehydrated gypsum. The first, either in the form of quicklime or slaked lime, served to neutralize the excess of acid in wine that had soured or was naturally sour; and the second, normally added before fermentation, served to clarify and improve the wine. Several Latin authors describe the practice of treating wine with lime or with partly dehydrated gypsum. For example, Columella explains in one place[508] how wine is treated with gypsum, and in another place[509] how it is treated with either gypsum or marble. In several places Pliny mentions the practice of treating wine with different calcium compounds. For example, he remarks that "the people of Africa reduce the acidity with gypsum, and in some parts with lime"[510] (*Africa gypso mitigat asperitatem nec non aliquibus partibus sui calce*). This shows clearly enough that the ancients used both lime and gypsum in the treatment of wine. Greek authors have less to say about this practice. Dioscorides[511] only once mentions the use of *gypsos* in the preparation of wine, though he writes at length about the different kinds of wine and the methods of preparing

[508] XII, 26.   [509] XII, 20.   [510] XIV, 120.
[511] V, 82 (Wellmann ed., V, 72).

COMMENTARY

them. However, two significant accounts are contained in the collection known as the *Geoponica*. One of these, ascribed to a certain Didymus, may be translated in part as follows: "The *gypsos* should be put into a broad vessel, and then the must should be poured on it so that it covers the *gypsos*. It should be shaken constantly and then left to stand, so that the coarser parts of the *gypsos* may sink to the bottom."[512] In the other account in the *Geoponica*[513] it is stated that when *gypsos* is added to wine, it makes the wine sharper at first, but in time this sharpness disappears. Apparently the practice of treating wine with gypsum or with lime was very common in antiquity. But Pliny shows that the practice was not always looked on with favor when he mentions wines treated with marble, gypsum, or lime, and asks in a characteristic manner: "Where is the man, however strong he may be, who has not stood in dread of them?"[514]

Though lime or marble was evidently added to wines in ancient times to reduce their acidity, it is very probable that partly dehydrated gypsum was the material ordinarily added to grape juice before fermentation. The ancient writers who give full accounts of the practice all seem to specify this material. The second account in the *Geoponica*, which states that the sharpness of the wine increased after treatment with *gypsos*, clearly shows that the wine was treated with an excess of gypsum and not with lime or any other neutralizing agent. In these accounts *gypsos* always seems to mean partly dehydrated gypsum, not any of the other substances the ancients included under this name.

It is known that very early in modern times partly dehydrated gypsum, which is now called plaster of Paris, was often used in the preparation of wines in various Mediterranean countries. This practice is very common today in certain of these countries, where it has probably been in continuous use since ancient times. In Greece gypsum is frequently added to wine, though this does not seem to have been true at the time of Theophrastus. Stephanides[515] shows the extent of this practice in modern Greece when he remarks that all the gypsum now mined on the island of Melos is used in the preparation of wine.

[512] VI, 18.   [513] VII, 12, 5.   [514] XXIII, 45.
[515] *The Mineralogy of Theophrastus*, p. 144.

The treatment of unfermented wine or must with partly dehydrated gypsum is a practice now commonly called "plastering." In this treatment a reaction occurs between the added calcium sulfate and the potassium bitartrate present in solution whereby calcium tartrate is precipitated and soluble tartaric acid and potassium sulfate are formed. The precipitated calcium tartrate then carries down various suspended impurities, thus greatly clarifying the must. The removal of the potassium bitartrate also makes the coloring matter more soluble, so that the color of the wine is improved. Moreover, the fermentation is rendered more rapid and complete, and the wine is said to keep better. In spite of these obvious advantages, however, the practice is somewhat objectionable, because potassium sulfate is left in solution in the wine. In some countries the addition of plaster of Paris is regarded as an adulteration of wine: either it is forbidden by law or a restriction is placed on the amount of potassium sulfate that may be present in the finished wine.

If the emendation in this passage is sound, as seems very likely, this remark of Theophrastus is the earliest known allusion to the practice of treating wine with gypsum.

67. And painters employ it for some parts of their art. Though finely ground calcium sulfate is a satisfactory white paint pigment, and chemical analyses show that it was sometimes used in antiquity for this purpose, at least in Egypt,[516] no ancient author seems to include it among the colors used for painting. Hence it was probably not used to any great extent as a true paint pigment but only in the preparation of a white ground for painting. That it was actually so used in ancient Greek times is shown by some analyses of Rhousopoulos,[517] who found that the white ground on painted Athenian lecythoi of the fifth century B.C. consisted of calcium sulfate.

67. and so do fullers, who sprinkle it on clothes.
In section 64 Theophrastus has mentioned the use of Samian earth and of Tymphaic earth or *gypsos* for treating clothes. The primary

---

[516] Lucas, *Ancient Egyptian Materials and Industries*, p. 399.
[517] Diergart, *Beiträge aus der Geschichte der Chemie dem Gedächtnis von Georg W. A. Kahlbaum*, p. 181.

COMMENTARY

purpose of treating garments with such earthy substances was probably to whiten the discolored cloth, but it is also possible that whole cloth was likewise treated with these earths in order to "weight" or stiffen it, just as certain kinds of cloth, especially silk, are weighted with inorganic substances at the present time.

67. *It seems to be far superior to other earths for taking impressions.*

This translation of ἀπομάγματα as "impressions" is supported by a passage in the *Causes of Plants*[518] which contains the phrase τὰ ἀπομάγματα τῶν δακτυλίων ("the impressions of signets"). Though this alludes specifically to impressions of signets or finger rings, in the present passage the word appears to have a more general significance and probably means impressions or molds in general. Obviously, the kind of *gypsos* used for such a purpose could only have been calcined gypsum.

68. *It is also clear from the following example that it has a fiery nature; for once a ship loaded with clothes was itself burnt when the clothes became wet and caught fire.*

Although this story has obviously been condensed, the meaning seems clear enough. The argument seems to be that the clothes carried on the ship had been treated with *gypsos*, a substance that generates heat on contact with water; that water somehow had come into contact with the *gypsos* on the clothes, so that heat was generated and the clothes set on fire; and that this caused a general fire that destroyed the vessel. It is unlikely that this story is true, because the sort of *gypsos* used for cleaning or whitening clothes was ordinarily a natural earth that did not generate heat on contact with water. Even if dehydrated gypsum had been used for whitening the clothes, no heat would have been generated on contact with water, since the gypsum would probably have been completely hydrated while the clothes were being cleaned. Moreover, even if dehydrated gypsum had been present in the clothes or as a separate cargo on board the ship, no fire could have been started because it does not generate enough heat on contact with water to ignite organic materials. Furthermore, even if the clothes

[518] VI, 19, 5.

had been treated with *gypsos* in the form of quicklime, which seems highly improbable, this too would have become hydrated in the process, so that this substance, though it actually can generate enough heat on contact with water to ignite organic materials, could not have caused the fire on the ship. This story is even less plausible than the stories about Kydias and Kallias in sections 53 and 59. It is possible, of course, that a ship with separate cargoes of quicklime and clothing once caught fire because the lime became wet, or that a cargo of clothing that had been treated with *gypsos* once caught fire from spontaneous combustion or some other cause. Thus the story may have had a real basis, but the true cause was not understood and a wrong explanation was given.

69. *Gypsos is also burnt in Phoenicia and in Syria, where it is fired in a furnace.*
In section 64 Theophrastus has mentioned that *gypsos* was made in Phoenicia and Syria by burning stones. Here he explains how this was done.

Schneider and Wimmer bracket the words καὶ καίοντες which occur in the manuscripts, since they seem redundant in addition to the main verb καίουσι. If καίοντες is kept, it could perhaps refer to the initial step of firing the stones, whereas the main verb would describe the whole process of burning them.

69. *Marbles especially are burnt, and also the more ordinary kinds of stones, while cow-manure is placed alongside the hardest ones to make them burn better and more quickly.*
It is important to note that marble was used to produce *gypsos*; this shows that quicklime, which is obtained when marble is subjected to intense heat, was one of the substances listed under *gypsos*. The simpler or more ordinary kinds of stones, if the reading ἁπλουστέρους is correct, probably consisted of limestone; this was the most abundant rock in ancient Phoenicia and Syria, but marble also occurred there.

Schneider has a most ingenious suggestion about ἁπλουστέρους; he thinks that the wording may originally have been καὶ ἁπλῶς τοὺς στερεωτάτους ("and in general the hardest stones"). The

COMMENTARY

article τούς is needed with στερεωτάτους, and the word μέν is really redundant. Some noun like βόλιτον ("manure") is needed as the object of παρατιθέντες, and ἕνεκα has been added to govern τοῦ θᾶττον καίεσθαι ("to make them burn more quickly"). The manuscripts and the Aldine edition merely have τὰ τοῦ, but Turnebus wisely changed τά to ἕνεκα. Furlanus preferred διὰ τό, but this is not as good. Schneider rightly added βόλιτον, which he chose because Pliny[519] mentions that manure was used for this purpose.

Eichholz[520] accepts ἕνεκα, but thinks that βόλιτον should come before παρατιθέντες and take the place of μέν, which is superfluous. This is a great improvement. He also accepts καὶ ἁπλῶς τοὺς στερεωτάτους, but prefers the following translation: "and absolutely the hardest limestone at that."

In his treatise *On Fire*,[521] Theophrastus alludes to the preparation of gypsos in Phoenicia in a way that indicates the use of a high temperature. If a high temperature was used, this is enough to show that quicklime, not partly dehydrated gypsum, was the substance produced in Phoenicia and Syria. Gypsum would not have been roasted at a high temperature; if the temperature is even as high as 200°C., gypsum is totally dehydrated and takes up water again too slowly to be useful for most purposes. Moreover, the firing of the stone in direct contact with the fuel also shows that the product was quicklime, not dehydrated gypsum. In roasting gypsum, the fuel is not allowed to come into contact with the mineral because it might reduce some of the calcium sulfate to calcium sulfide.

69. and stays hot for a very long time.
This is true of lime prepared in a kiln, not only because a high temperature is reached, but because the lime is such a poor conductor of heat.

69. it is pulverized like ashes.
Probably κονία does not mean lime in this context, as it does in sections 9 and 68, for that would imply that Theophrastus re-

---
[519] XXXVI, 182. The text reads: *In Syria durissimos ad id eligunt cocuuntque cum fimo bubulo, ut celerius urantur.*
[520] D. E. Eichholz, "A Curious Use of μέν," *Classical Review*, LXVI (1952), 144-45.
[521] Sec. 66.

garded this kind of *gypsos* as different from lime. Here it seems to mean ashes. In the same way, the translation "powdery" was used where κονία occurred in section 40, and τέφρα, a similar word, was translated as "ashes" in section 19.

69. *From this it seems clear that its nature is entirely due to fire.*

Since Theophrastus has completed his discussion according to the general principles that he announced at the beginning and systematically developed throughout the treatise, there is no need to assume, in spite of this abrupt ending, that it ever extended beyond its present length.

WORKS CITED IN THE COMMENTARY

CITATIONS of classical works in the Commentary list the title, except when an author is now represented by a single work, the number of the book, and the number of the chapter, section, or verse in the standard edition.

The following list of books by modern authors cited in the Commentary does not include standard dictionaries such as that of Liddell-Scott-Jones or encyclopedias such as that of Pauly-Wissowa. Also excluded are editions of the treatise *On Stones*, since these are listed in the Introduction.

BOOKS

ABRAHAM, H. *Asphalts and Allied Substances*, New York, 1945.
AGRICOLA, G. *De Natura Fossilium*, Basel, 1558.

BAILEY, K. C. *The Elder Pliny's Chapters on Chemical Subjects*, London, 1929-1932.
BAUER, M. H. *Edelsteinkunde*, Leipzig, 1932.
BECKMANN, J. *History of Inventions, Discoveries, and Origins*, London, 1846.
BERENDES, J. *Des Pedanios Dioskurides aus Anazarbos Arzneimittellehre*, Stuttgart, 1902.
BERTHELOT, M. *Archéologie et Histoire des Sciences*, Paris, 1906.
BLÜMNER, H. *Technologie und Terminologie der Gewerbe und Künste bei Griechen und Römern*, Leipzig, 1875-1887.
BOERHAAVE, H. *Elements of Chemistry*, trans. T. Dallowe, London, 1735.

DANA, J. D. *Manual of Mineralogy and Petrography*, New York, 1909.
———. *System of Mineralogy*, New York, 1909.
DAVIES, O. *Roman Mines in Europe*, Oxford, 1935.
DE BOODT, A. B. *Gemmarum et Lapidum Historia*, Leyden, 1647.
DE LAET, J. *De Gemmis et Lapidibus Libri Duo*, Leyden, 1647.
DIERGART, P. *Beiträge aus der Geschichte der Chemie dem Gedächtnis von Georg W. A. Kahlbaum*, Leipzig and Vienna, 1909.

FORBES, R. J. *Bitumen and Petroleum in Antiquity*, Leyden, 1936.
FRAZER, J. G. *Pausanias's Description of Greece*, London, 1913.
FURTWÄNGLER, A. *Die antiken Gemmen*, Leipzig and Berlin, 1900.

GIESECKE, W. *Antikes Geldwesen*, Leipzig, 1938.

HEAD, B. V. *Historia Numorum*, Oxford, 1911.
HICKSON, S. J. *An Introduction to the Study of Recent Corals*, Manchester, 1924.

BIBLIOGRAPHY

HILL, G. F. *Handbook of Greek and Roman Coins*, London, 1899.
HOEFER, F. *Histoire de la Chimie*, Paris, 1866-1869.
HOLM, A. *Beiträge zur Berichtigung der Karte des alten Siciliens*, Lübeck, 1866.
HOW, W. W. and WELLS, J. *A Commentary on Herodotus*, Oxford, 1912.
HULTSCH, F. *Griechische und römische Metrologie*, Berlin, 1882.
HYDE, W. W. *Roman Alpine Routes*, Philadelphia, 1935.

JAEGER, W. W. *Diokles von Karystos*, Berlin, 1938.

KING, C. W. *Natural History of Precious Stones and of the Precious Metals*, London, 1870.
KIRCHNER, J. *Prosopographia Attica*, Berlin, 1901.
KRAUS, E. H., and HOLDEN, E. F. *Gems and Gem Materials*, New York, 1925.

LACHMANN, K. *In T. Lucretii Cari De Rerum Natura Libros Commentarius*, Berlin, 1882.
LAGERCRANTZ, O. *Papyrus Graecus Holmiensis*, Uppsala, 1913.
LENZ, H. O. *Mineralogie der alten Griechen und Römer*, Gotha, 1861.
LUCAS, A. *Ancient Egyptian Materials and Industries*, 2nd ed., London, 1934, 3rd ed., London, 1948.

MACALISTER, R. A. S. *The Excavations of Gezer*, London, 1912.
MC INNES, W., DOWLING, D. B., and LEACH, W. W. *The Coal Resources of the World*, Toronto, 1913.
MEISSNER, B. *Babylonien und Assyrien*, Heidelberg, 1920-1925.
MELLOR, J. W. *A Comprehensive Treatise on Inorganic and Theoretical Chemistry* (Vol. III), London, 1923.
MIDGLEY, S. W. "Chemical Analysis of Ancient Athenian Pigments," unpublished senior thesis, Princeton University, 1936.
MOORE, N. F. *Ancient Mineralogy*, New York, 1859.

PARTINGTON, J. R. *Origins and Development of Applied Chemistry*, London, 1935.
PETERSON, C. L. "Egyptian Blue and Related Compounds," unpublished Master's thesis, The Ohio State University, 1950.
PETTUS, J. *The Laws of Art and Nature in Knowing, Judging, Assaying, Fining, Refining, and Inlarging the Bodies of Confin'd Metals*, London, 1683.

RAWLINSON, G. *The History of Herodotus*, London, 1858-1860.
RICKARD, T. A. *Man and Metals*, New York, 1932.
ROSE, T. K. *Metallurgy of Gold*, London, 1915.
ROSSIGNOL, J. *Les Métaux dans l'Antiquité*, Paris, 1863.

BIBLIOGRAPHY

SALMASIUS, C. *Plinianae Exercitationes*, Utrecht, 1689.
[SHEAR, T. L., editor and contributor] *Classical Studies Presented to Edward Capps on His Seventieth Birthday*, Princeton, 1936.
SINGER, C., and SIGERIST, H. E. *Essays on the History of Medicine Presented to Karl Sudhoff on the Occasion of His Seventieth Birthday*, Oxford, 1923.
STEPHANIDES, M. K. *The Mineralogy of Theophrastus* (in Greek), Athens, 1896.
STILLÉ, A., and MAISCH, J. M. *The National Dispensatory*, Philadelphia, 1880.
STILLMAN, J. M. *The Story of Early Chemistry*, New York, 1924.

THOMPSON, D'ARCY. *A Glossary of Greek Fishes* (St. Andrews University Publications, No. 45), Oxford, 1947.
TOD, M. N. *A Selection of Greek Historical Inscriptions*, Oxford, 1948.
TROWBRIDGE, MARY L. "Philological Studies in Ancient Glass," unpublished doctoral dissertation, University of Illinois, 1922.

PERIODICAL ARTICLES

BRÜCKL, K. "Die Minerallagerstätten von Ostafghanistan," *Neues Jahrbuch für Mineralogie, Geologie und Paläontologie*, LXXII, Abt. A (1936), 37-56.

CALEY, E. R. "Ancient Greek Pigments from the Agora," *Hesperia*, XIV (1945), 152-56.
———. "On the Prehistoric Use of Arsenical Copper in the Aegean Region," *Hesperia*, Supplement VIII (1949), 60-63.

DAVY, H. "Some Experiments and Observations on the Colours Used in Painting by the Ancients," *Philosophical Transactions of the Royal Society of London*, CV (1815), 97-124.

EICHHOLZ, D. E. "Aristotle's Theory of the Formation of Metals and Minerals," *Classical Quarterly*, XLIII (1949), 141-46.
———. "Bad Bronze Again," *Classical Review*, LIX (1945), 52.
———. "A Curious Use of μέν," *Classical Review*, LXVI (1952), 144-45.
———. "Theophrastus on ΠΟΡΟΣ," *Classical Review*, LVIII (1944), 18.

FOSTER, W. "Chemistry and Grecian Archaeology," *Journal of Chemical Education*, X (1933), 270-77.
———. "Grecian and Roman Stucco, Mortar and Glass," *Journal of Chemical Education*, XI (1934), 223-25.
FOUQUÉ, F. "Sur le Bleu Égyptien ou Vestorien," *Comptes Rendus Hebdomadaires des Séances de l'Académie des Sciences* (Paris), CVIII (1889), 325-27.

BIBLIOGRAPHY

GETTENS, R. J. "Lapis Lazuli and Ultramarine in Ancient Times," *Alumni* (*Revue du Cercle des Alumni des Fondations Scientifiques à Bruxelles*), XIX (1950), 342-57.

GOWLAND, W. "Arts of Working Metals in Japan," *Journal of the Institute of Metals*, IV (1910), 4-41.

———. "Silver in Roman and Earlier Times," *Archaeologia*, LXIX (1917-1918), 121-60.

HAMMER, J. "Der Feingehalt der griechischen und römischen Münzen," *Zeitschrift für Numismatik*, XXVI (1908), 1-144.

LAURIE, A. P., MC LINTOCK, W. F. P., and MILES, F. D. "Egyptian Blue," *Proceedings of the Royal Society of London*, LXXXIX A (1914), 418-29.

LEAF, W. "The Commerce of Sinope," *Journal of Hellenic Studies*, XXXVI (1916), 1-15.

MARYON, H. "Soldering and Welding in the Bronze and Early Iron Ages," *Technical Studies in the Field of the Fine Arts*, V (1936), 75-108.

RIDGEWAY, W. "How Far Could the Greeks Determine the Fineness of Gold and Silver Coins?" *Numismatic Chronicle*, Ser. 3, XV (1895), 104-109.

ROBERTS, W. C. "Alloys Used for Coinage," *Journal of the Society of Arts*, XXXII (1884), 881-91.

SCHMEISZER, C. "Bodenschätze und Bergbau Kleinasiens," *Zeitschrift für praktische Geologie*, XIV (1906), 186-96.

THOMSON, G. "Bad Bronze," *Classical Review*, LVIII (1944), 35-37.

WATSON, W. "Some Observations Relating to the Lyncurium of the Ancients," *Philosophical Transactions of the Royal Society of London*, LI (1759), 394-98.

WILSON, J. N. "Milowite—An Unusual Form of Silica," *Chemical Trade Journal*, XCVII (1935), 28; *Sands, Clays, and Minerals*, II, No. 3 (1935), 127-30.

GREEK INDEX

The names in this index are mainly those of places and persons and the varieties of stone or earth that are mentioned by Theophrastus. The numbers refer to the sections of the Greek text.

ἀδάμας, *adamas*, probably corundum, 19
Ἄθως, Athos, 64
Αἰγύπτιοι, Egyptians, 7, 24
Αἰγύπτιος (κύανος), Egyptian (*kyanos*), 55
Αἴγυπτος, Egypt, 6, 34, 55
αἱματῖτις, *haimatitis*, probably red jasper, 37
ἀκόνη, *akone*, whetstone, 44
ἀλαβαστρίτης, *alabastrites*, probably onyx marble, 6, 65
†ἄλπεις (ἀλιφάς), *alpeis* (*aliphas*, paints), 60
ἀμέθυσον, *amethyson*, amethyst, 30, 31
ἄμμος, *ammos*, sand, 21, 35, 40, 58, 59
ἀνθράκιον, *anthrakion*, a dark stone (see Commentary), 30, 33
ἄνθρακες, 12, 16, 39; ἄνθραξ, 18 (*anthrakes* [plural], coal, charcoal, 12, 16, 39; *anthrax*, 18)
ἄνθραξ, *anthrax*, a red stone (see Commentary), 8, 18, 19
ἄργυρος, 1, 4, 9, 39, 41, 46; ἄργυρος χυτός, 60 (*argyros*, silver, 1, 4, 9, 39, 41, 46; *argyros chytos*, quicksilver, 60)
Ἀρμενία, Armenia, 44
ἀρρενικόν, *arrhenicon*, orpiment, 40, 50, 51
Ἀρκαδία, Arcadia, 33
Ἀσία, Asia, 61
ἀστήρ, *aster*, star, 63
ἄσφαλτος, *asphaltos*, bitumen, 15
ἀχάτης, *achates*, a name for variegated stones, including agate, 31

Ἀχάτης ποταμός, Achates River, 31
Βαβυλώνιοι, Babylonians, 24
Βακτριανή, Bactriana, 34
βασανίζω, *basanizo*, to test, 45
Βῖναι, Binai, 12, 15
γῆ, ἣ ἕψεται, 49; ἡ τῷ χαλκῷ μιγνυμένη, 49; ἡ Μηλιάς, ἡ Κιμωλία, ἡ Σαμία, 62; ἡ Τυμφαϊκή, 62, 64 (*ge*, earth: Kimolian, Melian, Samian, 62; mixed with copper, 49; Tymphaic, 62, 64; which is boiled, 49)
γύψος, *gypsos*, a broader term than the English gypsum, 62, 64, 66

Δαρεῖος, Darius, 6
διάβαρος (διάβορος) λίθος, *diabaros* (*diaboros*, porous) stone, 20
Διοκλῆς, Diokles, 28
Δωριεῖς, Dorians, 37

Ἐλεφαντίνη (πόλις), Elephantine (city), 34
ἐλέφας, 6; ὀρυκτός, 37 (*elephas*, ivory, 6; dug up [fossil], 37)
Ἑλλάς, Hellas, Greece, 33, 67
Ἐρινεάς, Erineas, 15
Ἐρυθρὰ (θάλασσα), Red (Sea), 36
Ἔφεσος, Ephesos, 58

Ζεὺς (Διός), Zeus, 24

Ἠλεία, Eleia, Elis, 16
ἤλεκτρον, *electron*, amber, 16, 28, 29
ἡμιώβολος, *hemiobolos*, half-obol, 46
Ἡρακλεία (λίθος), Heraclean (stone), 4

GREEK INDEX

Ἡρακλῆς, Herakles, 25

Θῆβαι (αἱ ἐν Αἰγύπτῳ), Thebes (in Egypt), 6

Θηβαϊκὸς (λίθος), Theban (stone), 6

Θούριοι, Thourioi, 64

ἴασπις, *iaspis*, not jasper (see Commentary), 23, 27, 35

Ἰβερία, Iberia, 58

Ἰνδικὴ (χώρα), Indian (land), India, 36

Ἰνδικὸς (κάλαμος), Indian (reed), 38

ἰξός, *ixos*, birdlime, 49

ἰός, *ios*, verdigris, 57

Ἰταλία, Italy, 67

κάλαμος (Ἰνδικός), *kalamos*, reed (Indian), 38

Καλλίας (Ἀθηναῖος), Kallias (of Athens), 59

Καππαδοκία, Cappadocia, 52

Καππαδοκικὴ (μίλτος), Cappadocian (*miltos*), 52

Καρχηδών, *Charchedon*, Carthage, 18, 34

Κατάδουποι, *Katadoupoi*, First Cataract, 34

Κεία (μίλτος), (*miltos*) from Ceos, 52

κέραμος, *keramos*, pottery, 9

Κιλικία, Cilicia, 49

Κιμωλία (γῆ), Kimolian (earth), 62

κιννάβαρι, 58, 60; αὐτοφυές, τὸ κατ' ἐργασίαν, 58 (*kinnabari*, cinnabar, 58, 60; natural, prepared, 58)

κίσσηρις, 14, 19, 20, 22; ἡ ἐκ τοῦ ἀφροῦ, 19; ἡ ἐν Μήλῳ, 14, 21; ἡ λευκή, 22; ἡ μαλώδης, 22; ἡ μέλαινα ἐκ τοῦ ῥυακος, 22; ἡ ἐν Νισύρῳ, 21 (*kisseris*, pumice: 14, 19, 20, 22; black, 22; in Melos, 14, 21; in Nisyros, 21; *malodes*, 22; produced from foam, 19; white, 22)

κόλλυβος, *kollybos*, 46

Κόλχοι, Colchians, 58

κονία, *konia*, lime, also ashes or powdery ash, 9, 40, 68, 69

Κορίνθιος (λίθος), Corinthian (stone), 33

κουράλιον, *kouralion*, coral, 38

κριθή, *krithē*, 46

κρύσταλλος, *krystallos*, rock crystal, 30

κύανος, 31, 37, 40, 51, 55; ἄρρην, 31, 37; θῆλυς, 31; αὐτοφυής, 39, 55; σκευαστός, 55; Αἰγύπτιος (σκευαστός), Σκύθης, Κύπριος, 55; χυτός, 55 (*kyanos*, a variety of lapis lazuli, also a blue pigment: 31, 37, 40, 51, 55; Cyprian, Egyptian [manufactured], Scythian, 55; female, 31; fused, 55; male, 31, 37; manufactured, 55; native, 39, 55)

κυάνου φόρος, 55; χρώματα τέτταρα, 55 (tribute of *kyanos*, 55; four colors, 55)

Κυδίας, Kydias, 53

Κύπρος, Cyprus, 25, 27, 35, 64, 67

Κύπριος (κύανος), Cyprian (*kyanos*), 55

Λάμψακος, Lampsakos, 32

Λημνία (μίλτος), Lemnian (*miltos*), 52

Λιγυστική, *Ligystike*, Liguria, 16, 29

λίθοι, αἱ ἐκ τῆς Ἀσίας ἀγόμεναι, 61; οἱ ἐκ τῆς Βακτριανῆς, 35; αἱ ποικίλαι, 48; οἱ τίκτοντες, 5 (stones: from Asia, 61; from Bactriana, 35; variegated, 48; which give birth to young, 5)

λιθοκόλλητα, mosaics, 35

λίθος, ἡ βασανίζουσα τὸν χρυσόν, 45; ᾧ γλύφουσι τὰς σφραγίδας, 44; ἡ ἐν Λαμψάκῳ, 32; ὁ ἐν Σίφνῳ, 42; ὁ ἐν Τετράδι, 15; ὁ ἐν τοῖς Σκαπτησύλης μετάλλοις, 17; ἡ (ὁμοία) τῷ

GREEK INDEX

ἀργύρῳ, 41; ὁ περὶ Βίνας, 12, 15; ὁ περὶ Μίλητον, 19; ἡ τὸν σίδηρον ἄγουσα, 29 (stone: at Binai, 12, 15; at Miletus, 19; in Lampsakos, 32; in the mines of Skaptē Hylē, 17; in Siphnos, 42; in Tetras, 15; [like] silver, 41; that attracts iron, 29; that tests gold, 45; with which seals are carved, 44)
Λιπάρα, Lipara, 15
Λιπαραῖος (λίθος), Liparaean (stone), 14
λυγγούριον, 28, 31; θῆλυ, 31 (*lyngourion*, 28, 31; female, 31)
Λυδὴ (λίθος), Lydian (stone), 4

μαγνῆτις (λίθος), *magnetis*, Magnesian (stone), 41
μαλώδης (μηλώδης, μυλώδης) *malodes* (*melodes*, *mylodes*), 22
μαργαρίτης, *margarites*, pearl, 36
μάρμαρος, *marmaros*, marble, 9, 69
Μασσαλία, Massalia, Marseilles, 18, 34
μέλας (λίθος), black (stone), 7
Μηλιὰς (γῆ), Melian (earth), 62
Μῆλος, Melos, 14, 21, 63
Μίλητος, Miletus, 19
μίλτος, 40, 51, 52; ἡ Κεία, ἡ Λημνία, ἡ Σινωπικὴ (Καππαδοκική), 52; ἐρυθρά, ἔκλευκος, μέση, 53; αὐτάρκης, 53; ἡ ἐκ τῆς ὤχρας, 53; αὐτόματος, τεχνική, 55 (*miltos*, red ochre: 40, 51, 52; artificial, natural, 55; Ceian, Lemnian, Sinopic [Cappadocian], 52; light-colored, medium, red, 53; made from yellow ochre, 53; self-sufficient, 53)
μόλυβδος *molybdos*, lead, 56
μυλίαι (λίθοι), *myliai*, millstones, 9

Νίσυρος, Nisyros, 21
ξανθὴ (λίθος), *xanthe* (stone), probably yellow jasper, 37
ὀβελίσκος, *obeliskos*, obelisk, 24

οἶνος, *oinos*, wine, 67
Ὀλυμπία, Olympia, 16
ὄμφαξ, *omphax*, a green stone (see Commentary), 30
ὀνύχιον, *onychion*, a broader term than onyx (see Commentary), 31
ὄξος, *oxos*, vinegar, 56, 60
Ὀρχομενός, Orchomenos, 33
ὄστρειον, *ostreion*, oyster, 36

Πάριος (λίθος), Parian (stone), 6, 7
Πεντελικὸς (λίθος), Pentelic (stone), 6
Περραιβία, Perrhaibia, 64
πίννα, *pinna*, 36
πόρος, *poros*, travertine; here a kind of *poros* found in Egypt, 7
Πραξίβουλος (ἄρχων Ἀθήνῃσι), Praxiboulos (archon at Athens), 59
πρασῖτις, *prasitis*, a green stone (see Commentary), 37
πύελος, *pyelos*, sarcophagus, 6
πυρομάχοι (λίθοι), *pyromachoi*, fire-resisting (stones), 9

ῥύαξ, *rhyax*, lava stream, 22

Σαμία (γῆ), Samian (earth), 62
Σάμος, Samos, 63
σανδαράκη, *sandarake*, realgar, 40, 50, 51
σάπφειρος, *sappheiros*, a blue stone (see Commentary), 8, 23, 37
σάρδιον, 8, 23, 30; θῆλυ, ἄρσεν, 30 (*sardion*, a red stone [see Commentary], 8, 23, 30; female, male, 30)
σίδηρος, *sideros*, iron, 9, 28, 29, 43, 44
Σικελία, Sicily, 15, 22, 31
Σινωπικὴ (μίλτος), Sinopic (*miltos*), 52
Σινώπη, Sinope, 52
Σίφνος, Siphnos, 42
Σκαπτὴ Ὕλη, Skaptē Hylē, 17

GREEK INDEX

Σκύθης (κύανος), Scythian (*kyanos*), 55

σμάραγδος, 4, 8, 23, 24, 27, 35; σμάραγδος, ἴασπις, 27; ψευδὴς σμάραγδος, 25 (*smaragdos*, a green stone [see Commentary], 4, 8, 23, 24, 27, 35; false *smaragdos*, 25; *smaragdos* and *iaspis*, 27)

σπίνος, *spinos*, probably an asphaltic bitumen, 13

στατήρ, stater, 46

†στιρὰν (Ἄστυρα), *stiran* (Astyra), 32

Συήνη, Syene, 34

Συρία, Syria, 64, 69

σφραγίδιον, *sphragidion*, seal, 8, 18, 23, 24, 28, 30, 32

σφραγίς, *sphragis*, seal, 26, 44

τανός, *tanos*, a green stone (see Commentary), 25

τεταρτημόριον, *tetartemorion*, quarter-obol, 46

Τετράς, Tetras, 15

τέφρα, *tephra*, ashes, 19

Τμῶλος (ποταμός), Tmolos (river), 47

Τροιζήνιος (λίθος), Troezenian (stone), 33

τρύξ, *tryx*, wine-lees, here grape-residues, 57

Τυμφαία, Tymphaia, 64

Τυμφαϊκὴ (γῆ), Tymphaic (earth), 62, 64

Τύρος, *Tyros*, Tyre, 25

ὑαλοειδής, *hyaloeides*, glasslike stone (see Commentary), 30

ὑελῖτις, *hyelitis*, vitreous earth, 49

ὕελος, *hyelos*, glass, 49

Φοινίκη, Phoenicia, 55, 64, 67, 69

Χαλκηδών, Chalcedon, 25

χαλκός, 9, 28, 49; ἐρυθρός, 57 (*chalkos*, copper, 9, 28, 49; red, 57)

χερνίτης, *chernites*, apparently a variety of onyx marble, 6

Χῖος (λίθος), Chian (stone), 6, 7, 33

χρυσοκόλλα, *chrysokolla*, a green copper mineral, probably malachite as well as modern chrysocolla, 26, 39, 40, 51

χρυσός, *chrysos*, gold, 1, 4, 39, 45, 46

χρυσοῦς, *chrysous*, a gold piece, 18

Ψεφὼ (Ψεβώ), Psepho (Psebo), 34

ψιμύθιον, *psimythion*, white lead, 55, 56

ὤχρα, *ochra*, yellow ochre, 40, 51, 52, 53

INDEX TO TRANSLATION AND COMMENTARY

Achaia, a source of yellow ochre, 174
Achates, a variegated chalcedony or jasper, 52, 128-29
Achates River, 52, 129
Adamas, a very hard stone, 48, 91-92, 148
Aetites, eaglestone, 69
Agate, 127-29
Alabaster: called *onyx* by Pliny, 128; oriental, 72, 214
Alabastrites, probably Egyptian onyx marble, 46, 59, 72-73, 214
Alexander the Great, 52, 130
Alexandria, manufacture of blue frit at, 185
Alloys: assaying or testing of, 54, 150-56; copper, 164-67; gold, 54, 106, 151-53, 155-56; silver, 54, 152, 156
Alum, 207
Alyattes, 158
Amalgams, 204
Amber, 48, 51, 86, 111-13, 116-17
Amethyst, 51-52, 121-22
Ampelitis, an earth smeared on vines, 167
Amulets: coral, 141; eaglestones, 69; lapis lazuli, 127
Analyses and identifications of ancient materials, 171-72, 174, 177, 181, 189-90, 194
Anthracite, 86
Anthrakion, a very dark or black stone, 51, 52, 130-31
Anthrax, a red precious stone, 46, 48, 89-92, 130
Apion, 102
Arabic stone, 94
Arcadia, 52
Aristotle, 63, 64, 65-66, 159-60
Armenia, whetstone obtained from, 54, 149-50
Arrhenicum, Latin name for orpiment, 171
Arsenic sulfides, 171-72
Asbestos, 87-88
Ashes, 49, 60, 78, 92, 221-22
Asia, stones brought from, 58, 206-207
Asian stone, 207
Asphalt, 79, 80, 85
Assaying in antiquity, 150-59
Assian stone, 207
Assus, 207

Aster, a variety of Samian earth, 212-13
Astrion, a precious stone, 119
Astyra, 52, 129-30
Aswan: quarries of granite at, 72; site of ancient Syene, 132
Athens, identifications of pigments in excavations at, 173, 174, 177, 185, 190, 194
Athos, Mt., 59, 213
Attica: ochre of, used by Micon and Polygnotus, 175; source of yellow ochre, 173-74; white lead found in graves in, 189-90
Attractive power, stones which exhibit an, 46, 51, 67-68, 113, 117-18
Azurite, 105, 143-44, 183

Baal, 104
Babylonians, king of the, 50, 101
Bactria (or Bactriana), stones from, 52, 102, 133-34
Badakshan, the ancient source of lapis lazuli, 127, 184
Balearic Islands, a source of red ochre, 179
Barleycorn, an Attic weight, 154
Basalt, quarried in ancient Egypt, 75
Basanite, used for touchstones, 157
Belemnite, wrongly identified with *lyngourion*, 109-10
Beryl, 149, 217
Bina (or Binai), 47, 48, 80-81, 82, 85
Birdlime, 55, 167-69
Bithynia, the source of *spinos*, 81
Bithynians, in Thrace, 81
Bitumen and bituminous substances, 80-82, 85
Blue frit, 183-87
Blue vitriol, 207
Bones, found in the earth, 136
Borax, wrongly identified with *chrysocolla*, 105
Boz Dagh, the ancient Mt. Tmolos in Lydia, 157
Brass, ancient knowledge of, 164-65
Bronze, purification of, 167

Calcareous tufa, 74
Calcium compounds: carbonate, 67, 78; oxide, 210; sulfate, 210, 218

· 231 ·

INDEX

Caligula, the attempt of, to find gold in orpiment, 200
Cappadocia, a source of red ochre, 56, 175
Carnelian, 123
Carthage, ancient point of export for precious stones, 48, 52, 90-91, 104
Ceos, a source of red ochre, 56, 176-77
Ceramic ware, classed as stone, 161
Cerussa, Latin name for white lead, 187-91
Chalcedon, 50, 104
Chalcedony: banded, 127-28; green, 138; red, 122-23
Chalk, 210, 213
Charcoal, 47, 48, 85-86
Chernites, probably a variety of Egyptian onyx marble, 46, 73
Chian marble, 71-72
Chian stone, 46, 52, 71-72, 75-76, 130-31
Chios, a source of variegated marble, 71-72
Chrysoberyl, 138
Chrysocolla, modern meanings of, 105
Chrysocolla, Latin name for *Chrysokolla*, 105, 106
Chrysokolla: a green copper mineral, 50-51, 53, 56, 105, 143; a green pigment, 53, 174; an alloy for soldering gold, 106
Chrysoprase, 108, 121
Cilbian District, near Ephesos, a source of cinnabar, 195
Cilicia: source of a viscid earth, 55, 167-68; source of whetstone, 150
Cimolite, 209
Cinnabar, 57-58, 193-99, 203-204
Classification: of coral, 141; of earths, 169-70; of minerals, 129
Clay(s), 208, 209
Clay ironstone, nodules of, identified with *aetites*, 69
Clothes, cleaning or whitening of, 59, 60, 213, 218-219
Coal(s), 48, 81, 85-86
Coins, 90, 154-55
Colchis, a source of cinnabar, 57, 195-96
Collyrium, a variety of Samian earth, 212-13
Color: earths classified by, 56, 169-70; sex of stones determined by, 52, 124-25
Coloration: of glass by copper, 163; of water by *smaragdos*, 45-46, 50, 98-99
Combustion, 82-83, 93
Comum (Como), in Italy, source of a green stone, 145
Concretions, 68-69

Copper: 47, 51, 55, 57, 162-64, 165-67, 191-93; alloys of, 164-67; red, 57, 191
Copper acetate, 192-93
Copper carbonate, 105, 143, 174-75, 183
Copper salts, used in making imitation *smaragdos*, 98
Copper silicate, 105
Copperas, 207
Coral, 53, 140-41
Corinth: excavations at, 171-72, 189; source of an attractive stone, 52, 132
Corundum, 91, 147
Cosmetic, ancient use of white lead as, 190
Cow-manure, as a fuel for burning *gypsos*, 60, 220-21
Crete, a source of whetstones, 150
Croesus, 158
Crystals: of *anthrax*, 91; of colorless quartz, 121; of tourmaline, 109
Cuprite, 144
Cyitis, a stone containing embryo stones, 69
Cyprus: block or crystal of two colors found in, 51, 108-109; as a source of false *smaragdos*, 50, 104, of *gypsos*, 59, 60, 213, of *iaspis* and *smaragdos*, 52, of *kyanos*, 183

Daedalos, 203
Darius, 46
Dek, an island in Lake Tana, Ethiopia, 133
Demonesos, ancient source of copper minerals, 104, 183
Demostratus, 111, 115
Dendrachates, moss agate, 128
Density: of certain minerals, 53, 143; of pumice, 49, 95
Diabaros (*Diaboros*), porous stone, 49, 94
Diamond, 148
Didymus, 217
Diokles of Karystos, 51, 113, 201
Diorite (or Diorite-Gneiss), 75
Dioscorides, 94, 105, 107, 108, 111, 142, 143, 148, 167, 172, 178, 179, 181, 183, 189, 192, 207, 208, 209, 213
Dishes, 54, 146
Dolomite, 145
Dorians, 53

Eaglestone, 69
Earth: medicinal, 177-78; unusual, 55, 162-63; vitreous, 55, 161
Earthenware, brittleness of, explained, 78-79

INDEX

Earth-pits, 58, 205-206
Earths: artificial, 56-57, 181-93, 210; classification of, 55-56, 169-70; differences in, 45, 55, 169; natural, 58-59, 193-94, 208-210; occurrence of, 56, 58-59, 171-72, 173-74, 175-80, 183-84, 195-97, 205-206, 208-210, 213-14; origin or formation of, 45, 56, 58, 65-66, 170-71; uses of, 55, 56, 58, 59-60, 167-69, 174-75, 178, 185, 190-91, 193, 194, 210-11, 213, 215-19
Egypt, 46, 52, 72-73, 74-75
Egyptian kings, mention of records about, 50, 57, 100, 185-86
Egyptian materials and objects, 72-73, 74-75, 100-101, 127, 172, 184-85, 214
Electron, amber, 112
Electrostatic properties, of amber and *lyngourion*, 51, 111, 113, 117
Elephantine, 52, 132
Elephants, as a source of ivory, 135
Eleusis, materials for temple at, 202-203
Elis, a source of lignite, 48, 85-86
Emerald, 97, 99-100
Emery, 91, 147-49
Ephesos, cinnabar found near, 57, 195-97
Eretrian earth, 208
Erineas, a source of a combustible stone, 48, 84-85
Etna, lava of, 77
Euphranor, 181
Experiment: on blackening soapstone with oil, 146; on making white lead, 188

Felsite, red, 139
Ferric oxide: coloration of quartz by, 122, 139; in red ochre, 173, 177, 181, 182; in yellow ochre, 173
Fire: as a test of gold, 54, 150-51; formation of earths by, 45, 56, 65-66, 170-71; formation of pumice by, 49, 93; nature of *gypsos* due to, 60; origin of red ochre by, 57; power of resisting, 48, 92
First Cataract, 52, 132
Flower of Asian, or Assian, stone, 207
Fluorite, 108, 122, 138
Flux, limestone used as a, 77
Foam, supposed origin of one kind of pumice from, 49, 93
Fossil bones, 136
Fossil ivory, 53, 135-36
Fossil resin, 112-13
Fuel, use of mineral, in antiquity, 48, 86
Fullers, use of *gypsos* by, 60, 218-19

Galen, 125-26, 178, 207
Galena, 142-43
Garments, ancient, of asbestos, 88
Garnet, 89-90
Gassinade, a stone containing embryo stones, 69
Gaul, amber carried by trade routes through, 116
Gems, ancient engraved, 108, 119, 121, 122, 128
Gender of Greek names for stone and rock, 125-26
Generative power, in stones, 68-69
Geodes, as a source of quartz crystals, 76, 122
Georgia, a source of cinnabar, 196
Gezer, bronze from, 165
Gifts, sent to Egyptian kings, 50, 57, 101, 185-86
Glance pitch, 168-69
Glass: 55, 102, 103, 119-20, 161-62; volcanic, 83
Glass pastes, 119-20
Glass workers, 102, 103
Gold: 45, 46, 53, 54, 58, 63, 142, 150-53, 199-200, 203-204; alloys of, 54, 106, 151-53, 155-56; coins, 90; methods of gilding, 203-204; soldering of, 50, 106; testing of, 46, 53, 67-68, 151-58
Grain, as a unit of weight, 154
Granite, 72, 75
Grape-residues, 57, 191
Gypsos, a broad term that included gypsum, dehydrated gypsum, and lime, 58, 59-60, 210, 213-22
Gypsum, 72, 145, 210, 213, 214-19, 221

Haematites, Latin name that included our present hematite, 138
Haematitis, Latin name for *haimatitis*, 138
Haimatitis, probably red jasper, 53, 138-39
Half-obol, as a unit of weight, 55, 153-55
Hematite, 138-39, 177
Heraclea, in Pontus or Lydia, probable source of lodestone, 67
Heraclean stone, 46, 67-68, 118
Herakles, large green stone in temple of, 50, 103-104
Hermos River, 156-57
Hippias, 117
Hipponicos, 199
Hyacinth, 110
Hyaloeides, a glasslike stone, 51, 119-20

· 233 ·

INDEX

Hysteron proteron, use of grammatical construction called, 206

Iaspis, a generic name applied to various colored transparent or translucent stones, 50-52, 107-108
Iberia, a source of cinnabar, 57, 195-96
Iconium, ancient mines at, 176
Imitations: *iaspis*, 107; lapis lazuli, 137; precious stones in general, 98; *smaragdos*, 97-98
Impressions, use of *gypsos* for taking, 60, 219
India, 53, 134-35
Indian reed, petrified, 53, 142
Inscriptions, 177, 202-203
Iron, 46, 51, 54, 117-18
Iron mines, 56, 177
Ivory: fossil, 53, 135-36; a stone resembling, 46

Jacinth, 110
Jade, 108
Jasper: black, 158; green, 138; red, 139; yellow, 139-40

Kallias, 58, 198-99
Kaolin, 209
Kaolinite, 212
Karystos, a source of asbestos, 88
Khalki, a source of copper minerals, 104, 183
Kimolian earth, 58, 209
Kimolos, 209
Kollybos, a unit of weight, 55, 153-55
Krithē, a unit of weight, 55, 153-55
Kyanos: a blue pigment, 53, 56, 57, 126-27, 143-44, 175, 183-87; a blue precious stone, 52, 53, 126-27; Cyprian, 57, 183; Egyptian, 57, 183-87; Scythian, 57, 183-84; sex of, 52, 124, 126
Kydias, 56, 181-82

Labyrinth of Egypt, colossal statue in, 102
Lampsakos, gold mines at, 52, 129-30
Langurium, another name for *lyncurium*, 111-12, 115
Lapis lazuli, 126-27, 137, 183-84, 186
Lapis specularis, a term which apparently included mica and selenite, 119
Laurion, silver mines at, 142-43, 174, 198-99
Lava, 49, 95, 96
Lazurite, 126

Lead, manufacture of white lead from, 57, 187-89
Lecythoi, composition of white ground on, 218
Lemnian earth, 177-78
Lemnos: a source of red ochre, 177-78; visit of Galen to, 178
Leyden Papyrus X: fire test for gold described in, 150; mention of the touchstone in, 157; recipes in, for gilding and silvering, 204, for preparing gold solders, 106, for purifying metals, 167, for whitening copper, 166
Lignite, 81, 86, 88
Liguria, 48, 51, 86, 116
Lime: included under the name *gypsos*, 210, 220-21; mortar made from, 215-16; prepared by burning marble, 60, 78, 220; used for neutralizing wine, 216-17
Limestone, 67, 74-75, 77, 220
Lipara, 48, 84
Liparean stone, 48, 83-84, 131
Lipari Islands, a source of obsidian, 83
Liparite, 83
Lisbon (Olisipo), ancient point of export of garnets, 91
Lodestone, 67, 117-18
Lydia, 67-68, 158-59
Lydian stone, a name for the touchstone, 46, 67-68, 157
Lyncurium, Latin name for *lyngourion*, 109-12, 115
Lyngourion: amber or a particular variety of amber, 51, 52, 109-16, 124; sex of, 52, 124
Lynx, the supposed producer of *lyngourion*, 111-12, 113-16, 124

Magnesia, various minerals named after, 144
Magnesian stone, 53-54, 144-45
Magnetism, 117-19
Magnetite, 118, 145
Malachite, 100-101, 104, 105, 143-44, 174
Malodes, possibly a kind of pumice, 49, 96
Marble: burning of, to obtain lime, 60, 78, 220; Chian, 71-72; Egyptian onyx, 72-73, 74, 214; infusibility of, 47, 78; Parian, 70-71; Pentelic, 71; use of, in neutralizing wine, 216-17
Marcasite, 143, 145
Marpessos, Mt., a source of Parian marble, 70
Massalia (Marseilles), ancient point of export for precious stones, 48, 52, 90-91

INDEX

Medicinal uses: of Lemnian earth, 178; of pumice, 96; of verdigris, 193
Melian earth, 58-59, 208-209, 210-12
Melkart, the Tyrian Herakles, 104
Melos: a source of pumice, 48, 49, 84; the source of an earth, 208-209
Mercury, 203-205
Metals: alloys of, 54, 106, 151-56, 164-67; amalgams, 204; arsenic, 166; copper, 47, 51, 55, 57, 162-64, 165-67, 191-93; formation and origin of, 45, 63; gold, 45, 46, 53, 54, 58, 63, 142, 150-53, 199-200, 203-204; iron, 46, 51, 54, 117-18; lead, 57, 187-89; mercury, 203-205; minerals that resemble, 142; quicksilver, 58, 203-205; silver, 45, 47, 53, 54, 63, 142, 152-53, 156, 204; testing of, 46, 54-55, 67-68, 150-56
Metrodorus, 115
Micon, 175
Miletus: *anthrax* found at, 48, 91-92; deposits of emery near, 91
Millstones, 47, 117-18
Milowite, modern trade name for Melian earth, 209
Miltos, red ochre, 172-73, 179-80
Mineral substances, origin of, 45, 63-66
Mines: accidents in, 56, 176; at Binai, 47, 80; cinnabar, 57, 195-97; copper, 51, 56, 104-105, 183; gold, 51, 52, 56, 105, 130, 158-59; iron, 56, 177; lost work of Theophrastus on, 64; red ochre, 56, 176-77; at Scaptē Hylē, 48, 87; silver, 56, 58, 142, 198-99; *spinos* found in, 47, 81-82; unusual stones found in, 53, 144; yellow ochre, 56, 174
Mirrors, of obsidian, 52, 131
Miscarriage, prevented by amulets of eaglestone, 69
Modica, deposits of rock asphalt at, 85
Moonstone, 119
Mortar, made of lime or *gypsos*, 59-60, 215-16
Mosaics, stones used for, 52, 133-34
Mysia, a source of orpiment, 172

Names given to the same mineral substance in antiquity, 51, 74, 123-24
Naxos, stone of, 148-49
Nephrite, 108
Nisyros: abundance of millstone on, 77; pumice or volcanic ash on, 49, 95
North Sea, coasts of, a source of amber, 116
Obelisk of Zeus, 50

Obol, an Attic silver coin or weight, 154
Obrussa, a term used by Pliny to denote the fire test for gold, 150
Obsiana (*Obsidiana*), Latin names for obsidian, 84
Obsidian, 75, 83-84, 131
Ochre: red, 53, 56-57, 172-73, 175-83; yellow, 53, 56, 173-74, 175, 181-83
Odontolite, 136
Odor of stones, 47, 53, 143
Oil: blackening of stone with, 54, 146; burning of a stone with, 48, 87-88; quenches burning Thracian stone, 83
Olisipo (Lisbon), ancient point of export of garnets, 91
Olympia: 48, 85; temple of Zeus at, 74
Omphatitis, a stone probably similar to *omphax*, 120
Omphax, a green precious stone, 51, 120-21
Onychion, a generic term that included onyx, sardonyx, and agate, 52, 127-28
Onyx, 127-28
Onyx marble, 72-74, 214
Orchomenos, a source of *anthrakion*, 52, 130-31
Ore, smelting of, 77
Oriko, ancient ochre mines at, 176-77
Ormuz, a source of red ochre, 179
Orpiment, 53, 56, 171-72, 175; attempt to obtain gold from, 200
Oyster, 52

Pactolos River, 157
Painters, 56, 58, 60, 175, 181, 218
Painting, 59, 181, 190-91, 193, 218
Paints, 58, 202
Papyrus Holmiensis. See Stockholm Papyrus
Papyrus Leidensis. See Leyden Papyrus X
Paradoxes: ease of cleavage and hardness of stones, 54, 147; incombustibility of *anthrax*, 89-90
Parian marble, 46, 70-71, 74
Parnassos, Mt., calcareous tufa found at, 74
Paros, marble of, 46, 70-71, 74
Particles, size of, in pigments, in relation to color, 54, 184, 187
Pearl, 52, 134
Pentelic marble, 71
Pentelicos, Mt., 71
Perrhaibia, a source of *gypsos*, 59, 214
Persia: alleged source of lapis lazuli, 127; the source of *tanos*, 102

INDEX

Persian Gulf: a modern source of red ochre, 179; a source of pearls, 53, 134-35
Petrifaction, of objects placed in stone vessels, 67
Pharis, supposed name of a place, 211
Pharmakitis, an earth used for treating vines, 167
Phlogiston theory, 171
Phoenicia: *gypsos* made or used in, 59, 60, 214, 220-21; tribute sent from, 57, 185-186
Pigments: blue, 56, 57, 175, 183-87; green, 56, 57, 174, 193; red, 56-57, 57-58, 171-72, 175, 176-83, 193-99; relation between size of particles and color, 57, 184, 187; specimens found in archaeological excavations, 171-72, 173, 174, 177, 181, 185, 189-90, 194, 218; white, 57, 58-59, 187-91, 210-11, 218; yellow, 56, 171-72, 173-74, 175, 181-83
Pinna, a bivalve mollusk, 52, 134
Pitchstone, 131
Plasma, 100, 108
Plaster of Paris, 210, 213, 217-18
Plato, 63, 117, 118
Pliny, 67-68, 69, 71, 72, 81, 83, 84, 91, 92, 94, 96, 97-99, 102-103, 105-106, 107-15, 121, 122, 123, 124-25, 127-29, 131, 134, 136-37, 138, 139, 141, 142-43, 148-50, 152-53, 158, 173-74, 175, 177, 178-79, 181, 191-92, 193, 197, 200, 207, 216-17
Polygnotus, 175
Pompeiopolis, realgar mine at, 172
Pontic wax, 166
Pontus, a source of orpiment, 172
Pontus River, 80
Poros (*poros stone*), a name denoting certain soft calcareous rocks, 46, 73-75
Porous stone, 49, 94
Porphyry, 75
Potstone, 145-46
Pottery, resistance of, to fire, 47, 78-79
Prase, 100
Prasitis, an opaque green stone, 53, 137-38
Prasius, 138
Praxiboulos, archon at Athens, 58, 200-202
Precious stones, in antiquity, value of, 90
Prehnite, 120
Prince Islands, 104, 183
Procreation, power of, in stones, 46, 68-69
Psebo, 133
Psepho, 52, 132-33
Pumice, 48, 49, 83-84, 92-96

Puteoli, blue frit manufactured at, 185
Pyrargyrite, 144
Pyrite: associated with cinnabar, 199; in coal, 81; particles of, in *sappheiros*, 137
Pyrobitumen, 81, 85

Quarries, 46, 70-73
Quarter-obol, as a unit of weight, 55, 153-55
Quartz: 76, 97-98, 108, 121-22, 132, 134; green, 97, 120; purple, 121-22; red, 122-23; rose, 108; staining of, 98
Quicklime: called *gypsos*, 210, 220-21; mortar made from, 215-16; prepared from marble, 60, 78, 220; used for neutralizing wine, 216-17
Quicksilver, 58, 203-205

Ragusa, deposits of rock asphalt at, 85
Reagents, modern use of, in assaying gold or silver alloys, 151-52, 155-56
Realgar: 53, 56, 171-72; decomposed, 166; ironlike, 166
Recipes: for imitating precious stones, 98; for solders for gold, 106; for whitening copper, 166
Red felsite, 139
Red lead, 189
Red ochre, 53, 56-57, 172-73, 175-83
Red Sea (meaning the Persian Gulf), 53, 134-35
Resin, fossil, 112-13
Rhyolite, 83, 84
Roasting of ochre, 182-83
Rock asphalt, 80-85
Rock crystal: 51, 76, 121; engraved, or carved into large objects, 121; imitations of other stones made from, 98
Rubrica, a red earth, 177
Ruby, 89, 90
Russia, a source of fossil ivory, 135

Salts, 207
Samian earth, 58-59, 209, 210-11, 212-13
Samos, 59, 209
Sand: 49, 57-58, 199; use of, in mortars, 215
Sandstone, 72, 79
Santorin (Thera), 93
Sappheiros, dark blue lapis lazuli sprinkled with pyrite, 46, 50, 53, 126, 136-37
Sapphire, 136
Sapphirus, Latin name for *sappheiros*, 136-37

· 236 ·

INDEX

Sarcophagus: of limestone with petrified objects inside, 67; of *chernites*, 46, 73
Sard, 123
Sarda, 122, 125
Sardion: a red stone, 46, 50, 51-52, 122-23
Sardonyx, 128
Scaptē Hylē, a mining district in Thrace, 48, 87
Scicli, deposits of rock asphalt at, 85
Scoria, volcanic, 95
Scyros, a source of yellow ochre, 174
Seals: stone used for carving, 54, 147-49; stones used for, 46-47, 50-52, 122-23
Sepiolite, 209
Serapis, colossal statue of, composed of *smaragdus*, 102
Serpentine, 104
Sex in stones, 52, 124-25
Sicily, 48, 49, 83, 84, 85, 140
Siderite, 69
Sil, Latin name for yellow ochre, 174
Silver: 45, 47, 53, 54, 63, 142, 152-53, 156, 204; alloys of, 54, 152, 156; stone that resembles, 54, 145; testing of, 54, 150-56
Simonides, archon at Athens, 201
Sinai Peninsula, the source of copper ore for Egypt, 101
Sinope, export point for red ochre, 56, 178-79
Sinopic red ochre, 56, 178-80
Sinopis, Latin name for red ochre, 178-80
Siphnos: 54, 145; stone of, 54, 145-47
Slate, black, as a material for touchstones, 157
Smaragdos: counterfeits or imitations of, 98; false, 50, 101, 103, 104; a green precious or semiprecious stone, 45, 46, 50, 51, 52, 97-102, 108-109, 133; large, 50, 100-102
Smaragdus, Latin equivalent of *smaragdos*, 97-98, 102, 133
Smelting of ore, 77
Smyrna, a source of yellow jasper, 140
Soapstone, an experiment with, 146
Soldering of gold, 50, 106
Solidification of mineral substances, 65
Spain, as a source of cinnabar, 195-96
Spathi, ancient ochre mines at, 176
Sphragis: a medicinal earth, 177; seal stone, 108
Spilia, ancient quarries of Pentelic marble at, 71

Spinel, 89, 91-92
Spinos, a combustible mineral substance, 47, 81-82
Star, a variety of Samian earth, 59, 212-13
Stater: a gold coin, 90; a silver coin and standard of weight, 54, 152-55
Steatite, 145-47
Stockholm Papyrus: recipes for imitating precious stones, 98, 107; verdigris listed as a coloring material, 193
Stone quarries, 46, 70-73
Stone(s): black, 49, 83-84, 130-32; blue, 52, 53, 126-27, 136-37; combustible, 47-48, 80-81, 82-83, 85-88; cut for seals, 46-47, 50-52; engraved, 119, 121-23, 128; fire-resisting, 47, 76-77; fusible, 47; green, 50-51, 53, 97-103, 105-108, 120-21, 137-38; hard, 54, 91-92, 147-49; having the power of attraction, 46, 51, 67-68, 113, 116-18; having the power of procreation, 46, 68-69; imitation, 97-98, 107, 119-20, 137; incombustible, 48-50, 89-90, 92; occurrence of, 46, 47-48, 50-51, 52-53, 53-54, 54-55, 70-76, 80-82, 83-87, 90-91, 93, 95-96, 100-101, 127, 129, 132-33, 144-45, 148-49, 156-58; organic substances classed as, 51, 53, 111-13, 116, 141; origin or formation of, 45, 64-65; peculiarities in, 45, 53; properties of, 45-46, 50, 54, 67-69, 98-99, 147-48; pregnant, 68; purple, 51, 52, 121-22; red, 48, 52, 53, 89-91, 122-23, 138-39; remarkable, 50, 51, 52; sex of, 52, 124-25; small and rare, 46-47, 48, 50, 100; translucent or transparent, 46, 51-52, 107-108, 119-26; uses of, 46, 48, 52, 54, 70-72, 73-76, 85-86, 99, 106, 131, 145-47, 150, 220; variegated, 52, 55, 128-29, 132; yellow, 53, 139-40
Stories or illustrative statements: attempt of Caligula to obtain gold from orpiment, 200; discovery by Kallias of a method for refining cinnabar, 58, 198-202; discovery by Kydias of a method of making red ochre, 56, 181-82; fire on a ship caused by *gypsos*, 60, 219-20; huge green stones, 50, 100-102, 103-104; origin of *lyngourion*, 51, 111-12, 113-16; wonderful stone found in the gold mines at Lampsakos, 52, 129-30
Sudines, 115
Sulfates, manufacture of, in ancient times, 207

INDEX

Sulfide minerals, 142-43
Sulfur, pumice colored yellow by, 96
Syene (Aswan): granite quarried at, 72, 132; precious stones exported from, 52, 132
Syria, 59, 60, 168, 214, 220

Talc, 145
Tanos, a green stone, 50, 102-103
Taygetus, Mt., a source of whetstone, 150
Temple: of Apollo at Delphi, and of Zeus at Olympia, 74; at Eleusis, materials used in building, 202-203; of Herakles at Tyre, 50, 103-104
Testing of gold and silver, 46, 54-55, 67-68, 150-56
Tetartemorion, an Attic weight, 155
Tetras, the stone found at, 48, 84
Thales of Miletus, 117
Theban stone, 46, 72
Thebes in Egypt, 46, 72
Theophrastus: date of his treatise *On Stones*, 200-202; lost treatises of, 64, 142
Thera (Santorin), the source of floating pumice, 93
Thesprotians, 86
Thourioi, a place where *gypsos* was made, 59, 214
Thrace: Bithynians living in, 81; *spinos* found in, 81
Thracian stone, 81-83
Tmolos, Mt., 157-59
Tmolos River, 55, 156-57
Touch needles, 151-52
Touchstone, 46, 54-55, 67-68, 150-53, 155-59
Tourmaline, 109, 110
Trademark, cakes of Samian earth stamped with, 212
Travertine, 74
Tribute, of *kyanos*, sent to Egypt, 57, 185-86
Troezen, the source of an attractive stone, 52, 132
Troodos, Mt., on Cyprus, mines of asbestos at, 88
Tufa: calcareous, 74-75; volcanic, 95
Turquoise, 103, 136
Tusks, fossil, 135
Tymphaia, a source of *gypsos*, 59, 210, 214
Tymphaic earth, 58, 59, 210, 213
Tyre, large green stone at, 50, 103-104

Ural Mountains, a source of malachite, 100-101
Urine of the lynx, the supposed source of amber and *lyngourion*, 51, 111-12, 113-16

Verdigris: color of *prasitis*, 53; manufacture of, 57, 191-93; use of, as a medicine or a pigment, 193
Vestorius, manufacture of blue frit at Puteoli by, 185
Vinegar, use of, in isolating quicksilver from cinnabar, 58, 204, in making white lead, 57, 188-89
Vitruvius, 181, 185, 189, 195-97, 203-204
Vivianite, 136
Volcanic ash, 95
Volcanic glass, 83
Volcanic islands, 83, 84, 93, 95
Volcanic scoria, 95, 96
Volcanic tufa, 95
Volcanoes, craters of, 49

Water: effect of, on burning *spinos*, 47-48, 81, 82-83; formation of metals from, 45, 63
Wax, Pontic, 166
Weighting of cloth, 219
Weights, used in testing gold, 54-55, 153-55
Whetstone, 54, 149-50
White lead: manufacture of, 57, 187-89; not used in mural painting, 191; toxic nature of, 190; use of, as a cosmetic, 190
Whitening of clothes, 213, 218-19
Wine, treatment of, with lime, marble, or partly dehydrated gypsum, 216-18
Wood: attracted by *lyngourion*, 51, 113; not the only fuel used in antiquity, 86
Worms, protection of vines against, 55, 167-69

Xanthe, probably yellow jasper, 53, 139-40
Xuthos, probable Latin name for *xanthe*, 139

Yellow ochre, 53, 56, 173-74, 175, 181-83

Zenothemis, 115
Zeus: obelisk of, 50; temple of, at Olympia, made of *poros stone*, 74
Zircon, 110

Printed by Libri Plureos GmbH in Hamburg, Germany